Fuel for

Waste Materials in

Fuel for Free?

Waste Materials in Brickmaking

Kelvin Mason

Published by Intermediate Technology Publications Ltd
trading as Practical Action Publishing
Schumacher Centre for Technology and Development
Bourton on Dunsmore, Rugby
Warwickshire CV23 9QZ, UK
www.practicalactionpublishing.org

ISBN 9781853396250

© Kelvin Mason, 2007

First published in 2007

All rights reserved. Only pages 147–50 can be photocopied.
No other part of this publication may be reprinted or reproduced or utilized in any form or by any electronic, mechanical, or other means, now known or hereafter invented, including photocopying and recording, or in any information storage or retrieval system, without the written permission of the publishers.

A catalogue record for this book is available from the British Library.

The contributors have asserted their rights under the Copyright Designs and Patents Act 1988 to be identified as authors of their respective contributions.

Since 1974, Practical Action Publishing has published and disseminated books and information in support of international development work throughout the world. Practical Action Publishing (formerly ITDG Publishing) is a trading name of Intermediate Technology Publications Ltd (Company Reg. No. 1159018), the wholly owned publishing company of Intermediate Technology Development Group Ltd (working name Practical Action). Practical Action Publishing trades only in support of its parent charity objectives and any profits are covenanted back to Practical Action (Charity Reg. No. 247257, Group VAT Registration No. 880 9924 76).

Index preparation: Indexing Specialists (UK) Ltd
Typeset by S.J.I. Services
Printed by Replika Press

Contents

Foreword

Most of the millions of small-scale brick producers in developing countries use wood-fired clamps to fire their bricks. Clamps suit them, because they require no investment and can be built near any clay deposit whenever some bricks have been moulded. Even their sizes can be adjusted to the available quantities of dried bricks. A major drawback of these clamps is their low fuel efficiency, particularly if they are small. This is contributing to the depletion of wood resources in brick producing areas and threatening the environment. Beyond that, it is now threatening the livelihoods of small-scale brick producers themselves. The increasing scarcity of wood has motivated some authorities to altogether ban its use for activities such as brick production. Elsewhere, scarcity has increased its cost so much that it now exceeds half of the overall cost of producing the bricks. As a result, many small-scale brick producers have gone out of business or are at risk.

Both brick producers and development agencies have been looking for ways to resolve this fuel problem. The ideal option would be to produce fuelwood in a sustainable way, but this rarely happens. Another option is to increase the fuel efficiency of brick production, thus reducing the demand for wood. But, whilst some improvements can be made to clamps, the biggest gains in efficiency are acquired by moving towards fixed kilns. These do, however, require a substantial investment and changes to the ways of working, which often put them beyond the reach of small-scale producers. Whilst this book occasionally touches upon fuel efficiency and ways to increase it, that is not its major focus. Instead, it considers a third option, that is to substitute a large proportion of the fuelwood by residues.

Agriculture and industry in developing countries produce millions of tonnes of residues which generally are considered to be waste materials. Many of these have a substantial calorific value. They include, for instance, sawdust, bagasse, rice husks, coffee husks, cotton stalks, various straws, coconut shells, coal dust and ashes, and various types of dung. Whilst several of these residues could find other uses, these are often not exploited. Thus, they are left to rot, or are burned, or even dumped in waterways; these ways of disposal all negatively affect the environment, to a varying extent. Yet, they could be used in many production processes that require some form of energy, including brick making. Apart from substituting fuelwood, which is the major benefit, some types of waste can also be mixed into the brick clays, up to a degree. That will save clay, which is another scarce resource. The waste may also act as a flux, reducing the temperature hence the energy required to fire the bricks. Finally, it can reduce breakages. All of these effects have additional positive impacts on the environment.

Practical Action, formerly known as the Intermediate Technology Development Group, has been working with small-scale brick producers in developing countries for around 30 years. The fuel question has always been central to this work, in the context of improving the livelihoods of the producers and the sustainability of their ways of production. This book draws in the first place on Practical Action's experience of working with brick producers on fuel substitution in three particular countries: Peru, Sudan and Zimbabwe. These cases are set within a context of developments in this field world wide, and of overall environmental considerations which are increasingly coming to the fore. The book's author, Kelvin Mason, has worked with small-scale brick producers in half a dozen countries, for Practical Action as well as other agencies. He has previously written *Brick by Brick* (ITDG Publishing, 2001) which addresses the issue of fuel efficiency in more detail.

This book concludes that there is some real potential in using residues as a substitute for fuelwood in brick firing. It can have a significantly positive environmental impact, particularly in cases where waste is currently burned or left to rot anyway. The case studies presented, as well as experience from elsewhere, suggest that the use of residues as fuel is technically feasible and often cost-efficient. Proven technologies include the incorporation of fine waste in brick clays, spreading waste in voids strategically situated in clamps or kilns, and the use of low-pressure briquettes in firing tunnels. The use of waste can save large quantities of fuelwood, e.g. 75% in the Sudan.

But the experience so far is not uniformly positive. There are some potential drawbacks in the use of residues, and these need to be carefully assessed and managed. First of all, the use of residues could lead to less complete combustion, hence more pollution, than when wood is fired. It is therefore important to adapt the clamp to allow adequate air to access the fuel, to achieve near complete combustion. Another constraint could be transport: if the waste materials are located too far from brick making sites, any environmentally or economic gains in brick production could be reduced substantially by the burden of transport. Finally, using waste is likely to affect the quality of the bricks produced, either positively or negatively. In the latter case, the percentage of waste used should be limited to levels that guarantee the production of bricks of suitable quality.

The use of waste in brick production merits further dissemination. That will require various stakeholders to play their part. Authorities need to elaborate and adopt policies that make this possible, within the context of broader sustainable development. Development agencies need to support further research and testing to adapt clamps and kilns to locally available residues. And brick producers will have to contribute their local knowledge and skills as well as carrying some of the risk. This book will form a good starting point for all those involved in small-scale brick production in different capacities.

Paul Hassing
Deputy Director
Department of Environment and Water
The Netherlands Development Cooperation

Acknowledgements

I would of course like to thank the authors who contributed to this book: Saul Ramirez Atahui of Practical Action Peru and Emilio Mayorga of TEPERSAC, Lasten Mika of Practical Action Zimbabwe, Dr A. H. Hood who drafted the chapter on Sudan, plus Ray Austin and Otto Ruskulis who contributed to the European and global perspective. In addition, thanks to all those Practical Action staff and their brickmaker partners involved in the projects this book draws upon. Thanks too to the editors and others at Practical Action Publishing for all their efforts. Finn Arler from the department of Development and Planning in Aalborg University kindly allowed me to incorporate his work on resources into Fuel For Free?, so thanks to him also. Last but by no means least, thanks to Theo Schilderman of Practical Action UK, who nurtured the idea for this book for a number of years, and whose efforts finally permitted it all to be brought together.

List of figures

List of photos

List of tables

List of boxes

Abbreviations and units

N/mm^2 = Newton/square millimetre

1 N/mm^2 = 10.197 kilogram-force/square cm (kgf/cm^2)

1 kgf/cm^2 approx. equals 0.098 N/mm^2 approx. equals 14.22 lbf/in^2 (pound force per square inch)

1 GJ/tonne approx. equals 0.001 MJ/tonne approx. equals 0.239 calories/gram approx. equals 0.278 kWh/tonne

1 TPE equals 41868 MJ equals 11630 kWh equals 1.43 TCE

1 MJ/kg equals 239 cals/g equals 430 Btu/lb equals 278 kWh/tonne

toe = tonnes of oil equivalent

MPa = mega pascals

SD/SL£ is Sudanese dinars/pounds (COD has SUD as abbrev of Sudanese pounds)

feddan, unit of area, 1 hectare approx equals 2.38 feddas

MJ = megaJoules
kg = kilogram
TPE = tonnes of petrol equivalent
kWh = kilowatt hour
TCE = tonnes of coal equivalent
Cals/g = calories per gram
BTU/lb = British thermal unit per pound

CHAPTER 1
Small-scale brickmaking around the world

Kelvin Mason

> Up until the present day, development politicians have viewed 'poverty' as the problem and 'growth' as the solution. They have not yet admitted that that they have been largely working with a concept of poverty fashioned by the experience of commodity-based need in the northern hemisphere. With the less well off homo-economicus in mind, they have encouraged growth and often produced destitution by bringing multifarious cultures of frugality to ruin. For the culture of growth can only be erected on the ruins of frugality, and so destitution and dependence on commodities are its price... Whoever wishes to banish poverty must build on sufficiency. (Sachs, 1999)

Introduction

The Intermediate Technology Development Group (ITDG), recently renamed Practical Action, has been involved with small-scale brickmaking projects in so-called developing countries - I prefer to use the term 'majority world countries' - for the last 25 years at least. The principal objective of this work continues to be assisting brickmakers to secure more sustainable livelihoods from their enterprises. An associated objective is reducing the local and global environmental impacts of brickmaking. As an example of a local impact, consider the landscape degradation caused by excavating soil. The contribution to greenhouse gas emissions, specifically carbon dioxide, is an example of a global impact. This carbon dioxide contribution arises from both the burning of fuel to fire bricks and also, if wood is used as the fuel, the deforestation which that can cause. If woodlands are not sustainably managed, i.e. trees are not replanted on a complementary cycle, then burning wood means that natural sinks to absorb carbon dioxide are reduced. The same reasoning holds for any type of biomass. As is now almost universally accepted, the build-up of greenhouse gases due to human activity is transforming the life-giving greenhouse effect into the greenhouse problem, whereby the effect contributes to potentially catastrophic climate change via global warming.

With respect to both livelihoods and the environment, energy efficiency in brickmaking is critical. Not only does increasing energy efficiency serve to reduce brickmakers' fuel costs and hence increase their income, it also reduces the emissions of carbon dioxide and other pollutants per brick produced. If, as we will see, energy efficiency is combined with appropriate fuel substitution, sometimes known as co-firing, particularly when considered with respect to thermal power stations, then the beneficial effect on both income and the environment can be significantly enhanced. Efficiency and not wasting wastes, being *frugal* with resources, can therefore be the basis of truly sustainable livelihoods and is, in fact, the thesis driving this book. In Chapter 2 we will discuss the environmental aspects of brickmaking in much more detail.

In addition to objectives pertinent to brickmakers' livelihoods and a broad definition of environment, Practical Action has always been aware of the role that a secure supply of appropriate building materials, locally produced and affordable, can play in alleviating the shelter crisis that persists throughout much of the world. The story is not a simple one, however. If, for example, Practical Action is assisting brickmakers on the fringes of Zimbabwe's capital, Harare, to secure sustainable livelihoods in the prevailing conditions in 2005/06 then simply introducing the technology to make their operations more energy-efficient in isolation is unlikely to have the desired effect. Clearly, the 'market', i.e. the building materials needs of the mass of Harare's citizens, must be considered. As many people are living in so-called squatter camps or other forms of shelter that do not have official approval, then even making appropriate and affordable building materials available is not enough. For, who will be prepared to invest in a dwelling that they may be forced to pull down around their own ears at any time? Because the discourses on building materials and shelter are entangled and particular to place, Practical Action has integrated building materials and shelter programmes in a number of countries. These programmes address not only technology but also the social, environmental and political issues associated with the provision of housing and community facilities.

Considered globally, the scale and scope of the shelter crisis is extremely daunting. UN-HABITAT, the United Nations Human Settlements Programme, has been mandated by member states to improve the lives of at least 100 million slum dwellers by the year 2020. Though 100 million is an immense number of people to aim to assist in such a relatively short time, it is actually only some 10 per cent of those living in slums worldwide. UN-HABITAT forecasts that, unchecked, this number will increase threefold by 2050. So, by then we could be talking of 3 billion people who are inadequately housed in socially and environmentally unsustainable communities.

> As our towns and cities grow at unprecedented rates setting the social, political, cultural and environmental trends of the world, sustainable urbanisation is one of the most pressing challenges facing the global community in the 21st century. In 1950, one-third of the world's people lived in cities. Just 50 years later, this proportion has risen to one-half and will continue to grow to two-thirds, or 6 billion people, by 2050. Cities are now home to half of humankind. They are the hub for much national production and consumption – economic and social processes that generate wealth and opportunity. But they also create disease, crime, pollution and

poverty. In many cities, especially in developing countries, slum dwellers number more than 50 per cent of the population and have little or no access to shelter, water, and sanitation. (UN-HABITAT, 2003)

In 1989, having recently graduated from the Engineering Design and Appropriate Technology degree programme at the University of Warwick, I went to work in Zimbabwe. As a volunteer with International Cooperation for Development (ICD/CIIR), I began a professional involvement with brickmaking and building materials production that has lasted into the new millennium and continues as 2006 looms and I begin writing this book. The principal focus of my own efforts is the energy efficiency and environmental impact of small-scale building materials production. For most of the last 17 years I have been involved with Practical Action and its partners. In that time, I have had the privilege of working directly with building materials producers in Zimbabwe, Malawi, Kenya, Peru, Ecuador, Thailand and India. It is from these people that I draw the inspiration to make my small contribution to the enormous and vital challenge that confronts us all.

For the most part, the small-scale producers I have worked with understand not only the potential benefits of technological innovation to their own livelihoods, but are also very aware of the global environment and the duty that all of us have to protect it for both the present and also future generations. The Bruntland definition of sustainable development as 'development that meets the needs of the present without compromising the ability of future generations to meet their own needs' (WCED, 1987) is globally embedded, even if the practices

Photo 1.1: Dharavi, Mumbai, said to be the largest slum in Asia. Credit: Theo Schilderman.

of richer nations, particularly, continue to defy their rhetorical commitment. When members of a brickmaking cooperative in Zimbabwe, hard-pressed to sustain their own livelihoods, express their concern to reduce emissions of carbon dioxide for the sake of the now- and future-Earth, however, I can hold faith in humanity despite the worst efforts of multinational corporations and their political allies to ensure that this generation remains in poverty and the environmental future in jeopardy.

Susan George argues that ecology (the Logos) should be the guiding principle, superseding economy (the Nomos) which is merely the set of rules employed to for living up to that principle. She also highlights the conflict between sustainability and economic growth:

> (C)apitalism and environmental sustainability... are logically and conceptually incompatible. Two worldviews, the ecological and the economic, are locked in warfare, whether or not this war has yet been generally recognised. The outcome of the war will decide nothing less than the future of humanity and indeed whether or not humanity even has a future. (George, 2004)

There is a need to comprehend the difference between quantitative expansion (growth) and qualitative improvement (development), a difference which Herman Daly, Senior Economist in the Environment Department of the World Bank from 1988 to 1994, suggests is the defining feature of sustainable development (Daly, 1996): the core question is whether society has an economy or society *is* an economy. When economic development is measured in purely quantitative terms, such as Gross Domestic Product (GDP) per capita, it reflects nothing about sustainable livelihoods and environmental sustainability.

> Loss of resources, cultural depletion , negative social and environmental effects, reduction in quality of life – these ills can all be taking place, an entire region can be in decline, yet they are all negated by a simplistic economic measure figure that says economic life is good. (McDonough and Braungart, 2002)

Over the years, I have written numerous technical briefs on behalf of Practical Action and published books on the design, construction and operation of small-scale vertical shaft lime kilns (Mason, 1999) as well as on participatory technology development in brickmaking (Mason, 2001). I hope it will not depress the reader, or indeed myself, too deeply when I record that, since I began work in this sector, the overall shelter crisis has worsened substantially. Today's figures from UN-HABITAT paint a bleaker picture even than the introduction I wrote to *Brick by Brick* in 2001, which was bleak enough:

> However the problem is viewed, all statistics indicate that very little impact is being made on the world's huge housing deficit. As late as the 1980s, governments in some developing countries conceived rallying calls such as

'Shelter for all by the year 2000!'. In retrospect the exclamation mark appears sadly ironic... The slum areas that surround most Third World cities are a cause of suffering not only to inhabitants; they are also a threat to the nearby enclaves of the better off; and they are a blight on the economic development of the whole city... Just as the environment cannot be preserved with a quick fix, so the shelter crisis requires a long-term view and will ultimately have to involve rich people in developed countries making sacrifices in their standard of living [the quantity of their consumption rather than the quality of their well-being]... The trend towards a more open global economy [the neo-liberal project that is most often termed globalization] has meant that governments in the developing world have been obliged to be more market oriented and export driven. Public sector budgets have shrunk, and this has meant that support for the social sector has declined. (Mason, 2001)

As I write in November 2005, riots throughout France indicate that the problem of social polarization is not confined to so-called developing countries; excluded people living in deprived areas on the periphery of French cities do indeed pose 'a threat to the nearby enclaves of the better off'. Returning to the majority world context, though, it is evident that, despite the rhetorical commitment of UN member states, the work of Practical Action and the many other organizations striving to alleviate the shelter crisis is but a drop in a very large ocean. And yet, if there were genuine global political will, I am confident that the work Practical Action and others have done would stand us in good stead. Technologically, for example, I am certain that, resources permitting, we could help to improve the

Photo 1.2 Practical Action building materials and shelter work. Credit: Practical Action/Zul.

practice of small-scale brickmaking worldwide to the advantage of brickmakers, builders, the homeless and ill-housed, and the environment.

The challenge remains to disseminate the positive experiences of building materials and shelter programmes, such as those of Practical Action, on a mass scale. Too often, it seems, non-governmental organizations (NGOs) working at the grassroots are condemned to endlessly reproduce demonstration and pilot projects on the fringes while widespread worst-practice continues unabated at the core. In Zimbabwe in the 1990s, while the brickmakers whom Practical Action assisted strove to produce bricks that met the specifications for the urban market, we could only watch horrified and helpless as a new 'high-density' suburb was constructed from imported steel and cement. This mammoth project was funded by international donors and so, perforce, endorsed by the central government.

Not only was the market for locally produced building materials decimated virtually at a stroke, but the materials used were energy-intensive and, considering the transport energy alone, environmentally damaging. Moreover, the houses built were hot in summer, cold in winter and generally uncomfortable to inhabit. The architecture and site planning, meanwhile, can only be described as brutal: square boxes crammed together with not a tree left standing and no space for gardens or communal areas. This is not an example plucked from an inhumane colonial past, though the houses built are on a par with the worst of that era; this is a case of contemporary mass development practice. Lessons need to be learned. Appropriate technology is essential and, for me, that means the Schumacher ideal (Schumacher, 1973) as developed by Practical Action, i.e. building on small-scale, low-cost, environmentally friendly and non-violent local knowledge and skills via a dynamic and participative process.

With that conceptualization in mind, this book seeks to offer information gathered from grassroots projects that *could* be utilized on a mass scale. One of the technologies that has always attracted the interest of small-scale brickmakers and Practical Action alike is fuel substitution. Most small-scale brickmakers in the majority world use wood as their main fuel. Others, typically those based in urban or peri-urban areas whose production volume is higher, burn coal. In both instances, the price of fuel is continuously rising and can typically amount to 50 per cent of production costs. Moreover, secure and legitimate supplies of fuelwood are becoming scarce. People increasingly have to forage for fuelwood over long distances and perhaps risk breaking conservation laws, a situation I have witnessed first-hand in arid regions of Peru. Hence, brickmakers are universally quick to recognize the advantage of potential substitute fuels if they are readily available and affordable. Early in my experience as a fieldworker, I too perceived the possible benefits for both brickmakers and the environment.

Not only could burning waste save on the consumption of higher grade and perhaps scarce primary fuel, it could also be a valid means of disposal. Regardless of any calorific contribution to the process, toxic wastes or sewage could be rendered less harmful to human health by burning in a brick kiln, for example. In such cases, of course, the health implications for brickmakers would have to be seriously considered. Burning non-toxic wastes in brickmaking might be an environmentally valid and technologically straightforward means of disposal, however. Some wastes might act as fluxes, moreover, easing the process of vitrification in the clay brick and hence reducing primary fuel demand. Still other wastes might improve the moulding properties of the clay, reduce drying cracks in bricks, or improve the characteristics of the final product: enhancing appearance, increasing strength or inhibiting water absorption. Simpler yet, incorporating some wastes might reduce the volume of brick clay required, a consideration where clay reserves are scarce and could be conserved or where labour for extraction might be beneficially reduced. All in all, then, the substitution of wastes for fuel, for raw material, as a flux, or even utilizing brick firing as a means of waste disposal are all technologies that have potential livelihood and environmental benefits.

Through theory, reviewing appropriate literatures and, most of all, recording the everyday experience of Practical Action and our brickmaking partners, we will, in the course of this book, investigate the benefits and potential drawbacks of using a variety of wastes in brickmaking: Which wastes or residues can be safely and advantageously burned? Are sufficient quantities available to make technological exploration worthwhile in a particular setting? Do the wastes require processing prior to use? What percentage of the principal fuel or raw material can be advantageously replaced? What happens to the wastes if they are not used in brickmaking? And thence what is the net effect on the environment?

The history and extent of fired clay brickmaking

The history of moulding clay and firing it to yield bricks for building purposes has been traced from 5000 BCE through to the present day, and it makes very interesting and recommended reading (Campbell and Pryce, 2003). Conquering Roman legions, for example, utilized mobile kilns and so disseminated brickmaking widely throughout the empire. The reasons for firing clay bricks remain unchanged from pre-Roman times: to increase their resistance to water and weathering along with their strength, enabling more durable and extensive structures to be built. Brickmaking is, then, a very useful, well established and widespread technology, but one which retains a plethora of local variations worldwide. Small-scale brickmaking in particular has not yet been subject to any globalized process of standardization. From country to country and region to region, bricks, as well as floor and roof tiles, are formed in a variety of ways – moulded, pressed or extruded - into diverse shapes and sizes, dried naturally or mechanically, and fired in a range of clamp and kiln designs (with kilns generally taken to be permanent structures while clamps are constructed solely from the bricks to be fired) that accommodate anything from a few thousand to millions of bricks, and which utilize different fuels and variations thereof (e.g. fuelwood, wood pellets, wood chips, charcoal or sawdust).

Photo 1.3 Use of bricks in ancient times. Credit: Theo Schilderman

To get an idea about the nature, scale and scope of brickmaking in the world today we can compare Britain, which can perhaps be considered as something of a post-industrial economy, with India, a nation still substantially embroiled in industrialization. Although the available statistics are not always in directly comparable forms, the comparison is still enlightening. According to information from the Brick Development Association approximately 6,000 people are directly employed in brickmaking in Britain and many more work in ancillary industries (BDA, 2000). In terms of scale of enterprise ownership, only about 30 companies are involved in brickmaking, with just five being responsible for 84 per cent of production. The annual consumption of clay is around 8 million tonnes and this equates with 5.4 terawatts of energy consumption in drying and firing. Elsewhere the BDA translates this data into more comprehensible quantities, highlighting the energy and environmental advantages of using wastes as fuels in the process of so doing:

> The UK industry produces currently around 2.8 billion clay bricks per year. Over the last 20 years the energy requirement of their manufacture has been reduced by over 20%... Emissions from the firing processes have also been greatly reduced. The UK brick industry uses large quantities of waste materials: for example landfill gas is utilised for firing some bricks and in others colliery spoil, pfa [pulverized fly ash which will be explained and elaborated upon in Chapter 3] and blast furnace slag are used as fuel additives. (BDA, 2001)

The BDA claims that forecasts indicate that around 3 million additional homes will be built in the UK by 2020. Combined with the demand for bricks to renovate existing buildings, then, the market looks quite healthy, though, at 2.8 billion, 2001 saw the lowest demand for more than five years. Meanwhile, the industry is quite heavily regulated by a number of authorities concerned with resource conservation, energy efficiency and pollution. These authorities include the Department of Trade and Industry (DTI), the Department for the Environment, Food and Rural Affairs (DEFRA), the Environment Agency and the Local Authority. Emissions of hydrogen fluoride and particulates are limited by law, while newly built or redesigned plants are also limited in their permissible emissions of nitrogen oxides, hydrogen chloride and sulphur oxides. By 2010 the brick industry has undertaken to achieve a 10 per cent reduction in its specific energy consumption. This undertaking involves 28 manufacturers operating at 103 sites across the four nations of the UK.

A report by the Indian organization TERI (The Energy and Resource Institute) notes that bricks are one of the most important building materials in their nation (TERI, 2000). The Indian brick industry is reported to be the second largest in the world, dwarfed only by the burgeoning economic behemoth that is China. India has more than 100,000 brickmaking sites, producing about 140 billion bricks per year and consuming more than 24 million tons of coal and several million tonnes of biomass fuels, including of course fuelwood. Brickmaking is one of the largest employment generating industries in India, 'employing millions of workers'. To quote the TERI report further:

Photo 1.4 Large brickmaking site within the city of Pune, India. Credit: Theo Schilderman.

> Kilns are also notorious as highly polluting establishments, affecting not just flora and fauna, but also posing threats to human health. Higher energy costs and the inability of the industry to meet environmental standards has raised serious concerns about the survival and well-being of the industry.

Brickmaking is obviously key to economic development and the provision of shelter, creating jobs and producing an ever-popular and durable building material. When one considers the scale of brickmaking worldwide, however, particularly the growing significance of production in India and China, it is evident that, if livelihoods and the environment are to be sustainable, then everything possible must be done to achieve energy efficiency and maximize the use of appropriate wastes as fuel substitutes. These technological innovations are, of course, in addition to adequate and enforceable environmental regulation of the industry. At present the per capita carbon dioxide emissions of India and China are somewhere around a sixth of those of the UK's 7–15 tonnes. Imagine the environmental impact of only the brickmaking sector if these two industrializing giants take the same energy-intensive, environmentally destructive path to the same sort of economic development as the UK and other 'developed' nations.

While the global environmental impact of small-scale brickmaking in other regions, particularly the African continent, does not approach the same levels of significance as India and China, brickmakers face an intensifying struggle to sustain their livelihoods and the localized environmental effects are critical at that scale of concern. When bricks are underfired and therefore of less value because insufficient fuelwood is available, brickmakers lose out. When the local environment is plundered of every standing tree for miles around a brickworks, soil degradation and erosion are the likely result, often leading to the increased marginality of the land for food production. For a somewhat different range of reasons, then, fuel efficiency and utilizing wastes as fuels are equally significant concerns for brickmakers in more static, or indeed contracting, economies.

In the next chapter we will consider energy use and the environmental impact of brickmaking in more detail, beginning by defining what we mean by environment. Chapter 3 moves on to examine the technologies of fuel substitution and co-firing, considering a range of agricultural, industrial and other wastes and their potential for use in brickmaking, whether as fuels or otherwise. In Chapter 4, the use of coal-dust, coal-dust briquettes, waste oil, rice husks and sawdust in Peru is investigated. Chapter 5 looks at the case of Sudan and using a variety of agricultural wastes, including cow-dung and bagasse (a residue of sugar cane processing). The use of boiler waste in Zimbabwe is the subject of Chapter 6. Chapter 7 then offers a glimpse into possible alternative futures for brickmakers, reviewing the latest innovations in waste utilization in the industry worldwide. Considering the evidence from this review and the case studies in Chapters 4–6, particularly, Chapter 8 ponders the net environmental implications and the likely livelihood outcomes of the increasing use of wastes. This chapter also considers the political support, or lack of it, that might be mobilized in favour of energy efficiency and fuel substitution in the small-scale brickmaking sector around the world. Finally, the chapter attempts to draw everything together and conclude on possible courses of action for policymakers, fieldworkers and brickmakers.

CHAPTER 2

The environmental impact of using wastes

Kelvin Mason

> The fight against pollution [cannot] be successful if the patterns of production and consumption continue to be of a scale, a complexity, and a degree of violence which, as is becoming more and more apparent do not fit into the laws of the universe to which man is just as much subject as the rest of creation. (Schumacher, 1973)

Environment

We are working from the thesis that substituting wastes into the brickmaking process, either as fuels, soil conditioners or simply 'bulkerizers', can improve brickmakers' livelihoods and perhaps also reduce the environmental impact of brickmaking. We will look at the specific effect on livelihoods when we consider our case studies from around the world. And we will revisit the issue in a general sense in the concluding chapter, where we consider the future of using wastes in small-scale brickmaking. This chapter concentrates on the environmental impact of using wastes. Before we can consider the environmental impact of brickmaking and how that might be affected by using wastes, however, we should define what we mean by environment.

Albert Einstein, who was evidently right about a great deal, apparently said: 'The environment is everything that isn't me.' Aldo Leopold, the great US conservationist who worked and wrote in the first half of the 20th century, was an environmental thinker who believed that 'land', by which he essentially meant the environment, was a community that we should consider ourselves *a part of, rather than apart from* (Leopold, 1949). So, Leopold might have countered Einstein's definition with: 'The environment is everything *and* me', or even 'everything that *is* me', as some writers on the subject contest (Bullen and Whitehead, 2005; Whatmore, 2002; Latour, 1993)

In 2004 I was involved in a resource efficiency training programme with small-scale industries in Zambia. Participants came from a range of enterprise types and scales. One memorable course was attended by a farmer, a horticulturalist specializing in growing flowers and shrubs for city gardens, the owner of a small chemist shop, an entrepreneur whose main business was running a swimming pool but who also made bricks for use in extending the buildings at that facility, and the manageress of an extremely busy restaurant

Box 2.1 Introducing an environmental ethic

Discussion of what the environment is inevitably leads to questions about how we should consider it or how we should treat it. The fundamental cause of humanity's ongoing assault on the environment, many environmentalists agree, is the perception that the rest of nature exists exclusively to serve us (Croall and Rankin, 2000). This perception is reinforced by the Baconian creed, so dubbed after the 17th-century empiricist philosopher Francis Bacon. The Baconian creed asserts that scientific knowledge is technological power *over* nature. If we consider nature/environment as community, as Aldo Leopold would have it, then it is clear that service entails an obligation to contribute, to protect and to enhance rather than to exploit. Power, moreover, must surely be exercised responsibly.

The environmentalist philosopher J. Baird Callicott is one thinker who believes that 'ethics lie at the root of our environmental problems' (Palmer, 2001). As a means of solving these problems, Callicott judges the environmental stewardship ethic 'especially commendable' (Callicott, 1994). Philosophically, according to Callicott, the stewardship ethic invests non-human nature with intrinsic value, i.e. value independent of instrumental worth. It also means we have a duty of care to ensure that 'the earth's complement of species and inorganic natural appointments are not destroyed or degraded'. (Ibid.)

based in Lusaka's Soweto market, where among many other things one can buy the full range of Zambia's bountiful fruit and vegetable products, as well as second-hand clothes originating from seemingly every nation of western Europe. All these people were remarkably astute at identifying the 'non-product outputs' of their enterprises and quickly saw how they could save water, energy and materials, and hence improve their livelihoods. When it came to simply defining the environment, however, we were all a little stumped:

'Is it the air around us?' the chemist shop owner asked.
'Air and water,' the swimming pool owner proposed.
'And soil,' the horticulturist contributed.
'Yes,' the swimming pool owner agreed. 'But how about bricks which come from soil and water?'
'Animals,' the farmer thought, 'also birds and chickens.'
'My kitchen,' the caterer decided.
'Probably all those things,' I added lamely, wishing the training programme notes had included a definition as the basis for this discussion.
'All those things,' the farmer offered, wisely, 'and more.'

Indeed, many and much more. The participants were surely heading in many valid directions, however. It is one of the founding tenets of ecology that, as the famous naturalist John Muir put it, 'When we try to pick out anything by itself, we find it hitched to everything else in the universe' (Muir, 1992). The environmentalist Barry Commoner enshrined this principle as the first of his four laws of ecology: 'Everything is connected with everything else' (Commoner, 1971). One systematic definition of environment that will serve us practically in our investigation of fuel substitution in brickmaking can be gleaned from Environmental Impact Assessment (EIA) guidelines. Across Europe, if not more widely still, EIA guidelines tend towards an extremely comprehensive definition. EIA itself is a process for identifying, predicting and mitigating the adverse ecological and social effects of development projects and other human activities. In this process, environment is subject to a definition that takes into account scale and scope. With respect to environmental impacts, this means that both magnitude and range should be considered in space and time.

The local environment is the immediate vicinity (the caterer's kitchen, say). It is generally at this scale that health and safety legislation overlaps with environmental regulation. There is a logic to this overlap: it would be paradoxical to protect future generations through, say, sound policies on emissions of carbon dioxide while the current generation of workers were exposed to hazards that could injure them. Environmental impacts on the local environment include local pollution of air (smoke in the kitchen), water (supply to and drainage from the kitchen) and ecosystems. An ecosystem is the complex system of linkages – a network - between living things. So, ecosystems too can be conceived on a range of scales and scopes. Impacts can be confined to the local environment due to, say, the consumption of resources (landscape degradation post clay extraction for brickmaking, for want of another ready kitchen example). In addition, waste generation and disposal, noise and a range of other impacts may be assessed locally. They may, however, affect both the local and also other scales of consideration.

Photo 2.1 The local environment: Miriam's Restaurant. Credit: Lotte Reimer.

Environmental impacts on the regional scale are those that affect a larger area. We talk of regional rather than national because the political boundaries of nation-states have little relevance when considering environmental impacts; like the weather, pollution is no respecter of man-made borders. Consider the radioactive fallout from the 1986 Chernobyl nuclear power station disaster in Russia, an impact that is also an ongoing catastrophe for Belarus and Ukraine. Another example might be a factory discharging its liquid waste into a small stream and that stream carrying it to a river that disseminates the pollution effects regionally. Regional impacts, then, include flows of substances in the air, soil, water or ecosystems - from species to species, for example.

Major changes to natural areas may also have impacts at the regional level. Imagine, for instance, the effect on regional ecosystems of a seemingly local hydroelectric project that involves flooding a large area and changes the downstream flow of water through the seasons. Less obviously perhaps, consider the regional effect of a windfarm built in a national park. The windfarm may be quite benign in the natural environment, as well as being technologically wholly reversible in the long term, but what of its impact on the human environment? What, for instance, is the effect on all those people from all over the region, and indeed the world, who come to the national park to enjoy the beauty of its natural landscape, or on the livelihoods of those catering to such visitors?

With the recognition of problems such as ozone layer depletion, acid rain and, particularly, the human contribution to global warming and climate change, it is the global environment that is most likely to occupy news headlines these

Photo 2.2 What scale of impact from this Zimbabwean limeworks? Credit: Kelvin Mason.

days. Unfortunately, making guest appearances in the news does not seem to lead to concerted and effective human action to address these global problems. Stalling on the Kyoto Protocol, in itself an insufficient strategy to be effective against the human contribution to global warming, is a case in point. Apparently, we, the industrialized and industrializing nations, are, in general, too addicted to oil and consumption to forego an instant's pleasure for the sake of the wider world or future generations: there is an overwhelming lack of the popular support that could provide the political backbone governments need to make necessarily radical changes to the way we produce and consume. So, while governments, such as that of the UK, acknowledge the need for 'step changes' in resource efficiency in theory, permissible practice continues to be profligate (Moffat et al., 2001).

To illustrate the spatial and temporal nature of environmental problems on the global scale, the radioactive fallout from Chernobyl will continue to affect sheep on hill farms in distant North Wales for perhaps 30 years after the initial cataclysmic event. Other global-level effects include changes in sea currents and to marine life, the depletion of resources, including significantly fossil fuels and rainforests, desertification and loss of biodiversity. Though it is subject to a number of definitions, herein I consider biodiversity as the variety of life, plants, animals and micro-organisms, as well as their genes and also the ecosystems of which they are a part.

The loss or erosion of biodiveristy has increased dramatically in the modern era. Without doubt, this decline is due to human activities, particularly the destruction of natural habitats and the associated human consumption of organic resources, whether directly or indirectly. Ultimately, the loss of species, many of which, referring particularly to micro-organisms, we have not yet even identified, destabilizes ecosystems. Many commentators warn that the global ecosystem itself is therefore in grave danger of collapse (Leaky, 1996; Wilson, Dury and Chapman, 1999). Harvard professor of entomology and popular science writer Edward O. Wilson is popularly quoted as saying: 'If all mankind were to disappear, the world would regenerate back to the rich state of equilibrium that existed ten thousand years ago. If insects were to vanish, the environment would collapse into chaos.'

Apart from distinguishing between scales and scopes of environment, EIA guidelines also identify the physical as distinct from the socio-economic environment. It is perhaps the physical environment that springs to mind for most of us when the topic is raised. Air and the atmosphere, water resources, soil and geology, flora and fauna... These are aspects of environment with which we are probably most comfortable. But for EIA practitioners, the physical environment includes not only human beings but also our cultural heritage. To consider a definition of the physical environment, I find it useful to draw on Finn Arler's typology of resources (Arler, 2001). Starting from the principle of distributive justice and applying that to sustainable development, Arler proposes three overlapping categories of resources: critical, exchangeable and unique.

Critical resources are those that are important for all humanity; they determine survival and health. Good (enough) air and water quality as well as ecological services, such as the ozone layer and a globally beneficial greenhouse effect, are examples of critical resources. Exchangeable resources include consumer goods and, perhaps surprisingly, fossil fuels. The theory is that we can use fossil fuels now, provided it is not overly detrimental to the critical greenhouse effect, in exchange for developing, say, solar technologies that will meet the energy needs of future generations. Unique resources, as the name implies, are irreplaceable, largely irreparable, and, as a rule, non-exchangeable. They include historic buildings, rare species, works of art, and geographic areas that are aesthetically or biologically significant. It is geographic areas such as these that certain legislation in the British context, for example, seeks to protect by designating them as 'of outstanding natural beauty' or as 'sites of special scientific interest' (SSSI).

To return to our definition of environment, Arler's typology makes it apparent that a physical environment that includes cultural resources is a very wide and diverse sphere of concern. Unless circumstances were exceptional, I don't think development workers in majority world countries would consider the aesthetics of an area when assessing the potential environmental impacts of brickmaking, for instance; conventional economic considerations such as enterprise and employment creation are typically their prime considerations. Similarly, those development workers would not normally be aware of, or concerned by, rare species of micro-organisms in that area, species that perhaps thrive in

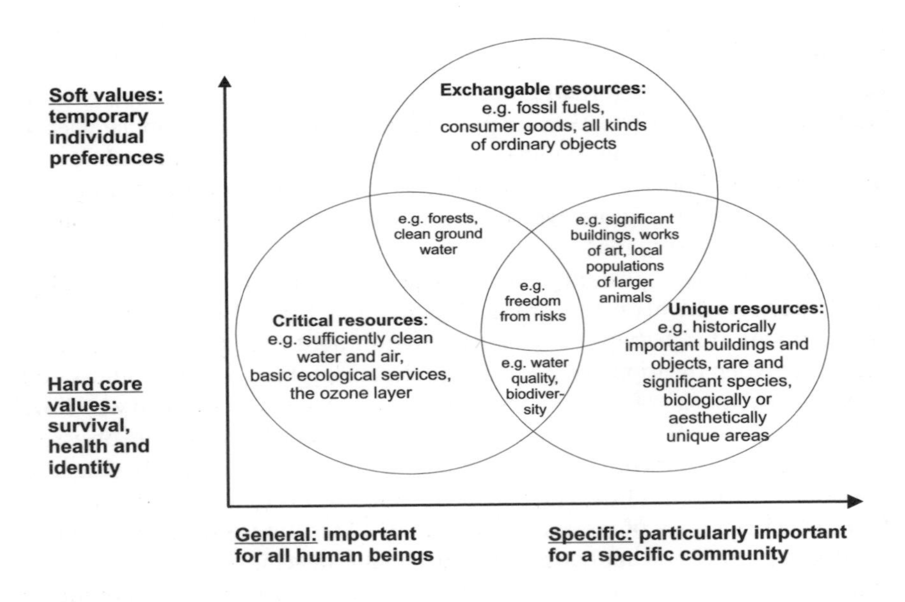

Figure 2.1 Arler's typology of resources (Source: Arler, 2001)

undisturbed topsoil. It may seem far-fetched to even raise the issue of considering some of these, seemingly more obscure, aspects of the environment with respect to fuel substitution in small-scale brickmaking. It may seem particularly 'cranky' in the light of, say, the magnitude of the shelter crisis I outlined in Chapter 1, with the evident desperate need both for bricks for housing and for employment to enhance livelihoods. As Florman reminds all of us who are champing at the bit of progress, however, it is the seemingly small and unconnected oversights and mistakes of one era that will add up to ecololgical crisis in a not too distant future (Florman, 1994).

The socio-economic environment, according to EIA guidelines, includes such considerations as employment, housing, services, health and education. So, when considering the impact of a proposed project, EIA practitioners would consider job creation, housing provision, improved access to essential services and so on as positive effects. They would then weigh these against, say, loss of species and landscape degradation. It will come as no great surprise to the reader, at least not the environmentalist reader, to learn that an EIA rarely stands in the way of any project that promises economic development (Christensen, Kørnøv and Nielsen, 2003a). EIA does, however, mean that the environment gets considered in a holistic way, and that perhaps the more damaging aspects of the project are modified. Moreover, the process of EIA continues to build our knowledge of environment, of the ecological connections between all things, and that may be its key contribution to sustainability (Christensen, Kørnøv and Nielsen, 2003b).

Some readers may have noted that the EIA definition of environment breaks down one of the traditional divisions of the concept, particularly as used by planners, i.e. that between the natural environment, which is not the result of human activity, and the built environment, which most definitely is. In EIA, and indeed Finn Arler's typology of resources, the natural and built environments mingle and merge. EIA's physical environment, which otherwise might be equated to the natural environment, contains cultural heritage. The socio-economic environment, furthermore, is not restricted to the environment built by human activity. Employment, for example, may be found in nature conservation, landscaping, or indeed performing an EIA! Trees, which are of course flora, can also be part of our cultural heritage; English readers may consider Sherwood Forest, the oak of England, upon which the sea-power of the empire depended, the Fortingall Yew, thought to be 5,000 years old and probably the oldest living thing in Europe, or the many other examples of trees of 'great historical or cultural importance' cited by the Tree Council (Stokes, 2004).

The division between the natural and human environment, indeed the Baconian schism between humanity and nature, is even more fundamentally challenged by the concept of hybridity, specifically socio-ecological hybridity (Latour, 1993; Whatmore, 2002). In fact, we have already touched on this concept without naming it. Historically important buildings in Arler's typology of resources have become socio-ecological hybrids, i.e. though man-made they come to be considered part of the natural environment to be conserved. Similarly, cultural heritage is considered as part of the physical environment in EIA. In a case study of the historical development of a cemetery, by way of a more specific example, the author claims that 'the non-human agency of trees has been enrolled into particular networks of environmental change and conservation' (Cloke and Jones, 2004). As 'socialized actors' the trees can be viewed as socio-ecological hybrids, i.e. they play a role in the social drama of environmental politics. Another way of thinking of this concept of socio-ecological hybridity is that who we are, our very identity, is based upon being a part of nature: we enrol the trees as actors in our social drama in the sense that they constitute a part of ourselves; they are not in themselves active, but we are inevitably activated by the them in us. In short, non-human nature and the social are not antitheses but syntheses (Whatmore, 2002). Such discussions echo the definitions of environment with which we began this chapter: everything that is *not* me versus everything *and* me versus everything that *is* me.

If the reader's conception of the environment is now not only more holistic than before but even includes an appreciation of the quite philosophical and academic concept of hybridity, then our consideration of the environmental impacts of brickmaking must surely be deepened and enhanced. The tree cut down for fuel may no longer be considered as mere biomass, but also a possible element of cultural heritage and even identity; the problem of deforestation can no longer be confined within the artificial boundaries of natural science and economy, it becomes a social and political issue as well. So, we will attempt to consider not only the physical magnitude of the environmental impacts of brickmaking, but also their socio-economic, political and cultural characteristics across a range of geographic scales and over time.

When considering such issues, an EIA-type conception of environment highlights the frequent conflicts of interest that arise. Often economic development, perceived as growth of GDP, is contrary to environmental protection. Moreover, social and environmental needs can easily conflict; reducing poverty today versus conserving resources for tomorrow, for example. Approached in a manner whereby spatial and temporal considerations are not taken into full account, 'making poverty history' - the popular campaign slogan which begs the question 'who made history poverty' - could mean that the future will be impoverished. Once again citing a popular Edward O. Wilson quote: 'It's obvious that the key problem facing humanity in the coming century is how to bring a better quality of life - for 8 billion or more people - without wrecking the environment entirely in the attempt.' And so, we will consider environmental impacts based on the principles of inter- and intra-generational justice enshrined in the Bruntland definition of sustainable development, drawing on the moral imperative of the environmental stewardship ethic, conscious of our duty of care.

Environmental impact

Having looked at the definition of environment at some length, we can now begin to consider the environmental impact of brickmaking specifically, and particularly how that impact might be affected by using wastes. Whatever the fuel used or the exact composition of the raw materials, making clay bricks involves a number of stages each with particular environmental impacts. We can list these stages as clay extraction, clay preparation, brick moulding, drying and firing. In environmental terms, we should also consider the transport of inputs to the site and of products from the site as stages of production. In order to begin to form an idea about the environmental impacts of brickmaking we can briefly consider these stages one by one. We will then move on to make an assessment of their impact.

Clay extraction obviously involves the excavation of soil with all that implies for the aesthetics of landscape, natural habits, future land use possibilities and so on. The extent of clay preparation depends on the nature of the soil. It may involve nothing more than mixing the soil with water prior to moulding. It could, however, mean sieving out stones, tempering (soaking) or weathering (stacking in heaps) to encourage homogenization, as well as grinding or milling the soil, particularly where tiles or more sophisticated products are the intended output. Moulding, pressing or extrusion is the process by which the brick is shaped. Most small-scale brickmakers in the majority world form bricks by slop or sand moulding. Sand moulding is a drier process than slop moulding with, as the name implies, sand employed as a releasing agent to prevent the drier clay sticking in the mould. In general, sand moulding produces a brick of higher quality in all respects: strength, water resistance, appearance, consistency of size and shape. Pressing machines are more commonly used in soil block production than in brickmaking. Extruders, usually motor driven but also human or animal powered in some cases, are used by some small- to medium-scale brickmakers.

Energy from fuel is obviously used in the firing and, if employed, artificial drying of bricks. Drying at the scales of production and in the dry-season geographical locations we are most concerned with is usually achieved by stacking bricks with spaces between them to take advantage of natural air flows and sunshine. In particularly sunny or dry and windy conditions, care has to be taken that bricks don't dry too rapidly and hence crack (Mason, 2000a). This is achieved by covering or shading brick stacks. Artificial drying is more common with larger-scale production and also in continuous firing processes, such as the Bull's Trench Kiln, which is used extensively in India. In such continuous processes, the waste heat from firing bricks is used to dry, or at least pre-heat, the next 'batch' of bricks. Firing can take place in a clamp or kiln and the two are distinguished by, as noted in Chapter 1, the latter having a permanent structure of some type. This structure can vary from the Scotch Kiln, a fairly rudimentary four-walled structure for the batch-firing of bricks, to the Bull's Trench and

Box 2.2 How big is an SME?

The question of enterprise scale is a somewhat vexed one. Generally, the definition of Small- and Medium-scale Enterprises (SMEs) employed in industrialized countries is quite different to that used in countries of the majority world. The European Commission (EC) defines enterprise scale in terms of turnover and 'headcount'. A medium-sized enterprise has a headcount of less than 250 employees, a small enterprise less than 50, and a micro enterprise less than 5 (European Commission, 2003). The EC has dropped a previous limitation on the percentage of an SME that can be owned by larger companies, thereby opening the door for large and transnational corporations to invest in the sector and presumably reap the benefits of any legislation favouring its development.

Taking Zambia as an example of a majority world country, the Government's Small Enterprises Development Act of 1996 also defines 'a small business enterprise' as one whose total investment and turnover do not exceed certain specified ceilings, albeit that these ceilings are very significantly lower than those set by the EC. The Zambians restrict the definition of such enterprises to those employing up to 30 people, and it is this simple measure that is most often referred to by NGOs working in the enterprise development sector, for example.

In development economics circles, *micro enterprises* are thought of as survival strategies that generate low income and meagre profits. They are founded on little or no capital, produce goods or services that have very low labour productivity, and typically operate in the informal sector, thereby avoiding taxation as well as environmental and health and safety legislation.

Vertical Shaft Brick Kiln (VSBK), which are both employed in continuous firing processes.

Because the structure provides thermal insulation, one advantage of kilns is less heat loss and hence less fuel use. In general, kilns are also easier to load and unload. Disadvantages include the capital cost and inflexibility. Despite the early ingenuity of the Romans, most kilns cannot be readily moved, whereas a clamp can be built wherever it is convenient for a brickmaker at a given time. Kilns, moreover, generally produce a set size of brick and the firing process is most efficient when they are full. This means it may be difficult for brickmakers to accommodate smaller and larger customer orders of bricks. An advantage of continuous firing processes lies in not being obliged to heat up the thermal mass of the kiln for each batch of bricks. Moreover, heat from bricks in the firing zone, which would normally be wasted in exhaust gases in a batch process, can be used to dry or preheat incoming bricks.

Ultimately, to answer what we could waggishly dub 'the burning question' about the environmental impact of using wastes in the firing of bricks, we must have at least a rudimentary understanding of combustion. For our purposes, then, combustion is the reaction between fuel and oxygen that results in heat being produced. The reaction also produces water and combustion products. The nature and quantity of heat, water and combustion products released depends not only on the type and nature of the fuel but also on the temperature at which combustion takes place and the quantity of oxygen available. The primary fuels

Photo 2.3 Landscape degradation after brick clay extraction. Credit: Kelvin Mason.

Photo 2.4 Vertical Shaft Brick Kiln in Nicaragua.
Credit: Martin Melendez, Grupo Sofonias en Nicaragua.

we are most concerned with, including wood and coal, are hydrocarbons - as are all fossil fuels, made up of varying proportions of carbon and hydrogen - and their combustion is actually a complex and frequently chaotic process. At a certain temperature, as it is heated towards its ignition point, a fuel undergoes pyrolysis, which means that it decomposes into an envelope of flammable gas plus liquid and solid products. The flammable gases then burn in a self-propagating fire.

'Complete combustion', an idealized notion, assumes that when a hydrocarbon burns in oxygen the reaction will yield carbon dioxide and water. The presence of other elements in the fuel, for example nitrogen, sulphur and iron, will result in oxides of these also being produced. A more likely scenario in reality, particularly in the rough and ready reality of brick firing, is 'incomplete combustion'. This means that there is insufficient oxygen for complete combustion and so, in addition to carbon dioxide and water, the reaction yields carbon monoxide. Incomplete combustion also produces larger amounts of polluting by-products. When we incorporate wastes into brick firing it is possible that there will be a trade-off between increasing pollution and lowering fuel costs. Evidently, we must find means of burning the waste efficiently, i.e. by means and in quantities that promote more complete combustion.

Transport of materials to and from the brickmaking site obviously has an environmental impact. Some brickworks are, at least initially, self-contained in terms of inputs, i.e. they are located at sites where good clay is abundant, sufficient water is available, and nearby trees supply the necessary fuel. This situation seldom lasts, however. In the absence of a programme of replanting, which is

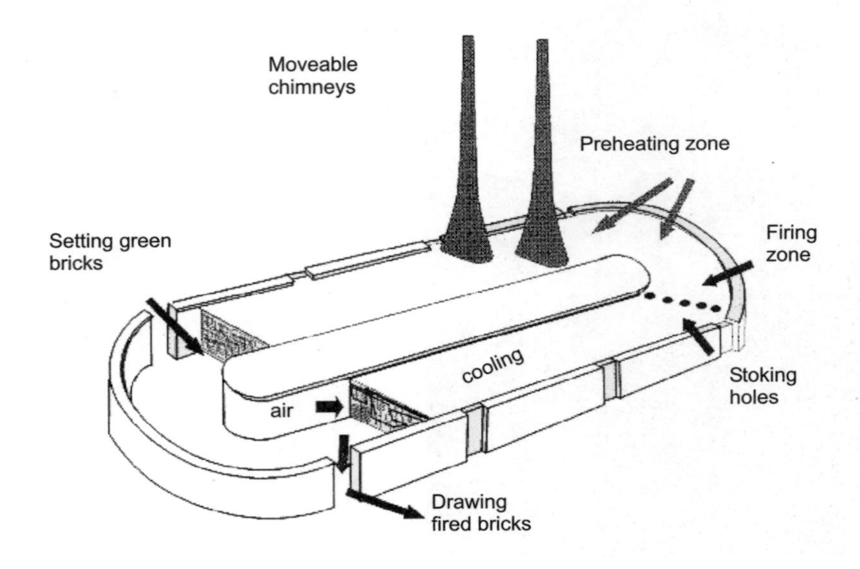

Figure 2.2 Bull's Trench Kiln (courtesy of basin)

generally the case, trees are consumed and brickworks end up buying in fuelwood or coal. This is delivered either by road or, if they are conveniently located, by rail. With respect to transport of products from the site, there is usually a significant environmental impact: bricks are seldom all used locally and most are transported away from the site by road or rail. The main implication for livelihoods and the environment of introducing wastes into brickmaking may well lie in transport to the site. While wastes may be freely or cheaply available, they may not be conveniently located close to brickmaking enterprises and delivery will involve both a financial and an environmental cost.

Environmental impact assessment

In order to consider the potential environmental impact of using wastes in small-scale brickmaking processes, I have set up a preliminary assessment form. The assessment considers not only the substitution of wastes for primary fuels, but also their use as soil conditioners and/or bulkerizers. It does *not*, however, cover the situation where firing bricks may be used as a means of toxic waste disposal because that technology does not feature in the case studies or future possibilities that will be discussed. Therefore, it is ultimately beyond the scope of this book. For now, I am assuming our wastes are neither too toxic nor too hazardous to be handled by small-scale brickmakers with little or no change to their established safety procedures. Because it appears to me to be quite well suited to our purpose, the assessment form used is based upon Danish guidelines for assessing 'bills and other Government proposals' (Denmark, 1995). I have assumed that we are considering a proposal to implement a range of waste substitution technologies on a nationwide basis in a certain country, perhaps one of our case study nations.

In carrying out this preliminary assessment, I have first assessed small-scale brickmaking on a nationwide basis *without* waste substitution technology, marking the form with an X in the appropriate box. Subsequently, I have considered impacts that would be likely to change *with* the introduction of substitution technology and highlighted these by doubling the X and marking them in bold, i.e. as **XX**. It is not that I presume the impact will change in each specific instance of technological innovation, but rather that I have highlighted where impacts may change in the wake of the implementation of a programme of fuel and raw materials substitution technologies on a national basis. The discussion that follows the presentation of the assessment form focuses on the impacts that may change and whether these changes would be beneficial to the environment, which according to the EIA definition we have adopted includes livelihoods.

The assessment is based only upon my informed opinion and the reader is encouraged to consider and contest my designation of effects. The discussion of the stages of brickmaking that has been presented in this chapter, however brief, should offer clues about areas of environmental impact. A blank assessment form is duplicated in Appendix 1. Readers are welcome to copy and use it to make their own assessment of our imagined project, which can then be compared with Figure 2.3. Whether or not they are concerned with brickmaking, readers may also wish to use copies of the assessment form to make provisional strategic assessments of other projects, real or imagined, on the national scale. Try, for example, comparing a proposal for nuclear power with an alternative proposal for wind power in the British context.

I will work through the assessment from the beginning in order, considering only the impacts that could change significantly due to the use of wastes. I have indicated with **XX** only that an impact may change. I have not speculated upon whether that change will be beneficial or detrimental to the environment; that is the subject of the discussion entered into henceforth.

Though they are probably of minor significance or insignificant, certain impacts on surface water could be adversely affected by the use of waste. Depending of course on the nature of the waste and its effect on the burning process, the discharge of organic substances could be increased to the detriment perhaps of ecosystems and natural habitats. Let us consider the example introducing animal dung into the clay mix as substitute for some part of the primary fuel. As a consequence, surface water may become more eutrophic, i.e. richer in nutrients and minerals. This could promote a proliferation of plant life, particularly algae, in water courses. This, in turn, reduces the dissolved oxygen content of the water and could result in the extinction of other organisms. Though, in specific cases of waste use, this might be an impact that would need

Table 2.1 National environmental impact of small scale brickmaking
Designation of environmental effects:
A=significant, B=should be examined, C=of minor significance, D=insignificant

	A	*B*	*C*	***D***
Is the proposal believed to cause a change in or effect:				
1. WATER				
1.1 Surface water				
– Discharges of organic substances, including toxic substances, into lakes & water courses?			**XX**	
– Discharge into coastal areas or marine waters?			X	
– Quantity of surface water or water level?				X
– Quality of salt water or freshwater?				**X**
– Natural ecosystems & habitats in salt or fresh water?			**XX**	
– Drinking water supply or reserves?				X
– Consumption/withdrawal of water?			X	
1.2 Groundwater				
– Percolation to groundwater?				X
– Groundwater quality?				X
– Quantity of groundwater?			X	
– Drinking water supply or reserves?				X
– Consumption/withdrawal of water?			X	

2. AIR				
– Emissions into air?	**XX**			
– Air quality (e.g. acid gases, particulate or toxic substances)?	**XX**			
– Obnoxious smells?			**XX**	
– Changes in precipitation quality?				X
3. CLIMATE				
– Emissions of greenhouse gases?	**XX**			
– Other factors, including deforestation, which may cause local or global changes in climate?	**XX**			
4. THE EARTH'S SURFACE & SOIL				
– Applicability or cultivation value of soil?		X		
– Percolation or accumulation of toxic or hazardous substances in the soil?				X
– Water or wind erosion?		X		
– Soil in the case of changes in groundwater level?				X
– The structure of the strata?			X	
5. FLORA & FAUNA, INCLUDING HABITATS & BIODIVERSITY				
– The number of wild plants or animals of any species or the distribution pattern of species?		X		
– The number or distribution pattern of rare or endangered species?		X		
– Import or export of new species, including GMOs?				X
– Quality or quantity of habitats for fish & wildlife?		X		
– Structure of function of natural ecosystems?		X		
– Vulnerable natural or uncultivated areas (e.g. bogs, heaths, uncultivated dry meadows, salt marshes, swamps and coastal meadows, watercourses, lakes, humid permanent grasslands and coasts)?		X		
– The reproduction or natural patterns of movement or migration of fish & wildlife species?			X	
– Cultivation methods or land use in the agricultural or forestry sectors?		X		
– Fisheries, catches or the methods applied in deep-sea or freshwater fishing?				X
– Open-air activities or traffic in the countryside which may affect the flora & fauna or cause wear & tear on the vegetation?			**XX**	
6. LANDSCAPES				
– The total area or the land use within areas used?	**XX**			
– Geological processes such as soil migration and water erosion?		X		
– Geological structures in the landscape, e.g. river valleys, ridges & coastal structures?			X	
– Permanent restrictions on land use which reduce the				

future possibilities of use of the open land?	**XX**			
– The extent or appearance of archaeological or historical sites, or other material assets?			X	
7. OTHER RESOURCES				
– Cultivation, cutting, catching or use of renewable resources, e.g. trees, fish or wildlife?	**XX**			
– Exploitation or use of non-renewable resources such as fossil fuels, minerals, raw material (sand, clay)?	**XX**			
8. WASTE				
– Wastes, residues or quantities of waste disposed of, incinerated, destroyed or recycled?		**XX**		
– Treatment of waste or its application on land?			**XX**	
9. HISTORICAL BUILDINGS				
– Buildings with architectural, cultural or historical value and with possibilities of preservation and restoration?				X
– Buildings or historical monuments which require repair because of a change of the groundwater level or air pollution?				X
10. PUBLIC HEALTH & WELL-BEING				
– Acute &/or long-term health risk in connection with food, drinking water, soil, air, noise, or handling of hazardous or toxic substances?			X	
– Risk associated with exposure to noise?				X
– Recreational experiences & facilities, including changes in the physical appearance of landscapes, natural or uncultivated areas?		X		
– The function & environment of towns, including green areas & recreational facilities?				X
– Aesthetic values or visual experiences (e.g. scenery, urban environment or monuments)?		X		
11. PRODUCTION, HANDLING OR TRANSPORT OF HAZARDOUS OR TOXIC SUBSTANCES				
– Risk of fire, explosions, breakdowns or accidents & emissions?			X	
– Risk of leaks of environmentally alien or genetically engineered organisms?				**XX**
– Risks associated with electromagnetic fields?				X
– Risk of radioactive leaks?				X
– Risk of breakdowns or accidents during transport of substances of materials?			**XX**	
– Other effects related to the security and safety of the population (e.g. traffic accidents, leaks)?			X	

to be promoted to category B, 'should be examined', I propose that in general the significance will not change from its 'minor' designation.

Depending on the nature of the waste and its effect on the burning process, emissions into air and air quality could be affected. Overall, I assume that wastes will not burn as completely as the principal fuels used in brickmaking; experience suggests that wastes tend to be more difficult to burn. Recall that incomplete combustion produces larger quantities of polluting by-products. Burning wastes may therefore mean more smoke, particulates and carbon monoxide. With respect to the local environment, I expect that some wastes may also increase obnoxious smells. Overall, burning wastes has the potential to worsen the already significant environmental impact of brick firing on air, increasing pollution. To jump ahead of ourselves, the way we burn wastes will obviously be crucial and we must aim to approach the ideal of complete combustion.

Assessed on our imagined national scale, brickmaking has a significant impact on the emission of greenhouse gases, specifically via the carbon dioxide produced by combustion. The use of wastes as fuel substitutes may directly reduce or increase this significance. If the burning process is maintained at a comparable level of efficacy, then it will require a similar input of fuel energy, i.e, a similar number of Joules or Watt-hours. Some fuels emit less carbon dioxide per unit of energy produced than others. Natural gas, for example, is a much 'cleaner' burning fuel than wood in this respect. So, the net effect of introducing wastes into the brick firing process will depend exactly on which primary fuel is replaced by which waste and in what proportion. In terms only of carbon dioxide emissions in the immediate term, for instance, it looks like a good idea to replace wood as a primary fuel with as high a proportion of diesel oil as possible. On the other hand, under the same proviso, it looks like a bad idea to replace coal with sawdust. If similar levels of combustion can be attained, burning sawdust is evidently the carbon dioxide emission-equivalent of burning wood.

Table 2.2 Carbon dioxide emissions

Fuel	*CO_2 (kg/GJ)*
Wood	109.6
Peat	106.0
Lignite	101.2
Hard coal	94.6
Fuel oil	77.4
Diesel	74.1
Crude oil	73.3
Kerosene	71.5
Gasoline	69.3
Refinery gas	66.7
Liquid petroleum gas	63.1
Natural gas	56.1

The above picture is of course incomplete. Specifically, it pays no attention to the effect of burning wood, or any biomass (in this context, plant material used as fuel) which is sustainably produced. Burning biomass is often taken as being 'carbon neutral' because the amount of carbon dioxide emitted when the biomass is burned cannot, even with complete combustion, be more than the amount it has absorbed during its growing life. Being carbon neutral when burned does not, however, make biomass a renewable or sustainable energy source. Only if new biomass is grown to replace that harvested and burned can the cycle continue sustainably. Otherwise, burning biomass reduces the available 'sink' for carbon dioxide and contributes to the greenhouse problem. So, if wood comes from a sustainably managed source, then it is better to burn wood than natural gas, at least from the point of view of carbon dioxide emissions. Moreover, if biomass derives from waste that would otherwise rot, thereby emitting its carbon dioxide content to the atmosphere anyway, albeit somewhat more slowly, then the emissions that would otherwise have come from the proportion of primary fuel replaced will be 'saved'.

Evidently, assessing the environmental impact of greenhouse gas emissions due to using wastes in brickmaking is going to be a complex process that will be highly dependent on the specific proposal. And we have not yet considered the technological practicalities, the economic realities and livelihood implications, or other aspects of sustainability. Taking these one at a time for a particular scenario: burning natural gas, indeed any gas, is not a technological possibility for most brickmaking SMEs. This is strongly related to the economic reality, where economic reality means the economic circumstances that currently exist rather than any permanent or essential condition. Even if brickmaking SMEs could manage natural gas burning technology, most could not afford the necessary hardware. Moreover, the gas itself would be either unavailable or unaffordable, or both. Current economic reality means there is no market cost on carbon emissions that would make high carbon emitting fuels more expensive. In economics terms, the externalities of burning such fuels are not accounted for and, contrary to the principles of sustainable development, the negative effect will be felt mainly by future generations. For the time being, then, unsustainable supplies of wood are likely to be a very much cheaper option than natural gas for the vast majority of brickmakers. The other aspect of sustainability to be considered is the finite nature of all fossil fuels, including natural gas. Treating this criterion in isolation would indicate that sustainably produced biomass was the only fuel choice for brickmakers.

Moving on to 'other factors' that may cause climate change, I had in mind deforestation specifically when assessing small-scale brickmaking. Many, perhaps the majority, of small-scale brickmakers with whom we are concerned burn wood from unsustainable sources. Trees, as we discussed, are sinks for carbon dioxide and so act to mitigate the greenhouse problem and thence climate change. By this logic, using wastes instead of timber from unsustainable sources as fuels could have a significantly beneficial impact on climate.

Considering 'flora and fauna', including habitats and biodiversity, I propose that the import of wastes by road could change the environmental impact. This will be the case in instances where primary fuel, say locally harvested wood or coal delivered by rail, were to be replaced by a waste product that necessitated delivery by truck. Obviously, the change could equally be in the opposite direction: if coal delivered by truck could be replaced by a locally produced waste or one that was delivered by rail, for example.

I pondered long and hard over whether the use of wastes could have an effect on the impact of brickmaking on landscapes, particularly on the total area of land used and permanent restriction on future use. I decided that if wastes were massively and extensively employed as bulkerizers, replacing huge quantities of soil across our imagined nation, then there would be changes to landscape impacts. To be specific, the area of land used by brickworks would decrease and so consequently would the area on which there were permanent or at least long-term restrictions on use. Though the environmental impact of using waste in this way would be beneficial to the landscape element of environment, in practice I doubt whether bulkerizers could be used on a scale where the change in land use was significant.

When the impact on 'other resources' is assessed, if tree felling in woodlands that are not sustainably managed can be reduced by the use of waste as a fuel substitute, then these resources are conserved. Similarly, fossil fuel resources can be conserved if, for example, a proportion of the coal used in a brick firing process can be replaced by waste. Considering the use of wastes as bulkerizers again, soil resources, which can be considered non-renewable within a certain timeframe, could conceivably be conserved. Notice that, given a long enough timeframe, soil, peat and fossil fuels may be renewable. Unless we conserve what is available and find alternatives, however, this timeframe is not one that is of much practical interest to humanity and the survival of our species.

Assessing the impact on 'waste', using waste as a fuel in brickmaking can be a way of incinerating it, i.e. disposing of a problem. The positive impact on local, and perhaps regional, environments might therefore be considerable. A waste dump, for example, may cause local pollution with an obnoxious smell and be literally 'a blot on the landscape'. In addition, it may well be a health hazard. Pollution on the regional scale, moreover, may be caused by harmful substances leeching out from the dump and polluting watercourses. Burning such waste is environmentally beneficial in these regards, but must of course be balanced against the consequent impacts in a different local environment perhaps, that wherein the brickworks operates, as well as globally. I have not highlighted 'treatment of waste or its application on land' from the perspective of burning wastes as a means of disposal. Rather, I considered the case of where animal dung was used as a fuel rather than as a fertilizer cum soil conditioner. If the soil were being deprived of fertility and the binding humus it needed, then using such waste as fuel could have a net-negative environmental impact.

The transport of, say, agricultural wastes to a brickmaking site for use as fuel could conceivably increase the risk of introducing an 'environmentally alien' organism considerably. There have been numerous recorded cases where such an event has meant disaster for the local and even regional environment. When alien species are introduced to an ecosystem they will likely disrupt the balance, perhaps introducing disease, degrading habitats and/or reducing biodiversity. Thus, an insect or even micro-organism transported from one (bio) region to another in a load of waste could have a long-term negative environmental impact much more significant than the immediate impact of brickmaking. Finally, I have assumed that the import of wastes will have an impact on transport and hence on breakdowns and accidents. In general, I assume using wastes will mean increased transport and hence increased risks in these regards.

In summary, the areas in which critical changes should definitely be assessed when introducing wastes into the brickmaking process appear to be:

- deforestation (carbon dioxide sinks and resource conservation);
- emissions of carbon dioxide;
- emissions to air and air quality;
- waste incinerated or otherwise disposed of;
- flora and fauna, including habitats and biodiversity (increased traffic).

Based on my knowledge of the brickmaking sector, I have listed the concerns above in something like a rough order of priority. Traffic is a problem not only because an increased number of trucks delivering waste may increase damage to flora and fauna and perhaps habitat and biodiversity. If wastes are delivered by road or rail, there will also be an increase in the emission of carbon dioxide due to burning the fuel that powers the trucks and trains. As we have seen, the cost of this externality is not accounted for. In certain circumstances, then, it may be financially beneficial for brickmakers to burn waste while the impact on the environment of so doing is wholly negative. Finally in this chapter, I should stress that this strategic EIA will only serve as a guide when we consider possibilities for the use of wastes in brickmaking. It does not obviate the need for a full EIA of individual projects. This observation is also true in the general sense: not every project for which a positive overall national assessment has been made would be environmentally beneficial, for example.

CHAPTER 3
Fuel choice and the potential of wastes

Otto Ruskulis

Although it is possible to describe brickmaking in general terms, the details and nuanced practices in play at individual sites are almost impossible to capture. The type of clay soil available largely dictates some of these practices. Methods of winning, and perhaps milling, tempering, mixing and using additives, vary from site to site, as do moulding, drying and firing practices. In addition to soil type, variation here depends on a multiplicity of factors, including tradition, resources and climatic conditions. The type of fuel used also varies from site to site, within clusters and even, depending largely on availability, within an enterprise. Distribution of the fuel in the clamp or kiln and whether it, or a proportion of it, is incorporated into the bricks, is also a variable. In some kilns, notably down-draught kilns, bricks are fired indirectly, utilizing the heat in exhaust gases without the fuel coming in contact with the bricks. Such burning methods are generally associated with a good-looking, higher quality product. On the other hand, one of the most efficient forms of fuel use is to embody it in the clay matrix so that the fuel is in intimate contact with the brick.

Before moving on to consider fuel choice and the possible substitution of brickmaking materials with wastes in detail, let us briefly consider some of the properties of clays and bricks. In the process, we will also touch upon the effects that adding residues and wastes could have on these properties. The mechanical or compressive strength of the fired brick is not normally a critical consideration for 'common' bricks used in normal construction applications. It is vital, by contrast, for products such as engineering bricks, which may be used in high load-bearing applications. Small-scale brickmakers in the majority world tend to produce, at best, common bricks. The specification of 'common' varies from country to country. In general, though, it is the lowest quality of brick recognized in official building standards. That said, the bricks made by many small-scale producers do not meet – or even need to meet – this baseline specification. In Zimbabwe, for example, the produce of most artisanal brickmakers is categorized as farm bricks, which in the main do not reach the compressive strength standards required for commons.

Nevertheless, Zimbabwean farm bricks and their international equivalents satisfy the technical requirements of a market niche that is significant, at least in terms of the number of customers. Dried clay is by itself quite a strong material, having a typical compressive strength of between 2 and 7 N/mm^2. If such clay

Photo 3.1 Rammed earth in Kenya. Credit: Practical Action/Neil Cooper.

can be kept dry by appropriate architectural design, then it is adequate for many construction applications. Structurally, clay can be used as rammed earth (pisé de terre), pressed soil blocks or adobe, which has straw added to reduce shrinkage and bind the clay. The main reason for firing clay bricks is to make them more resistant to water and weathering rather than to increase their strength.

The increase in strength on firing is a quite welcome side-effect, however. Typically, the compressive strength of common bricks is in the 20 to 40 N/mm^2 range. A farm brick would fall somewhere between the strength of an unfired clay brick and a brick classed as common. Most often, small-scale brickmakers would be aiming to just meet the requirements of the market. They would not want to produce bricks of high compressive strength that demanded a high input of fuel energy unless there was a premium payment involved. But if brickmakers cut costs too much, they risk producing bricks that perform no better than moulded clay. Using such bricks in architectural circumstances where they are expected to perform a load-bearing function in wet or abrasive conditions could be catastrophic.

Generally, when organic materials such as sawdust or coffee husks are added to the bricks before firing, mechanical strength is reduced. Adding too much of such a material as a fuel substitute can make the bricks too porous and friable: brittle and tending to crumble. It is this limitation, rather than lack of compressive strength, that makes the bricks unsuited to purpose. The amount of organic material that can be embodied in clay bricks as a fuel substitute is thus limited. With wastes that act as fluxes, conversely, compressive strength may be increased. If such an increase is not required, the addition of a flux may mean that, as an alternative, bricks can be fired at a lower temperature, saving fuel while maintaining acceptable properties.

The green, or unfired, strength of the brick is a consideration. Normally, bricks that have been moulded are strong enough to handle and subsequently stack prior to firing. It is the clay in the soil that is responsible for this behaviour, making particles in the material matrix cohere. Adding wastes or residues to bricks can reduce green strength and make handling a problem. Rice husks, for example, are large relative to the micro-structure of clay soil. The amount of rice husk that can be added may therefore be limited by its adverse effect on green strength notwithstanding its beneficial effect as a fuel. One way of getting beyond this limit is to grind rice husks before they are incorporated into brick clays. Rice husks can, of course, be burned in unlimited proportions if they are used as a non-embodied fuel. The limitation then becomes getting sufficient oxygen into the combustion process. Rice husks are high in silica and tend to pack together with little space for air to flow. Consequently, they are quite difficult to burn.

The density of common bricks is typically in the range 1,800 to 2,500 kg/m^3. When particulate fuel is included in the bricks this burns away and leaves pores, so the density of the final product is reduced. Bricks that are lighter in weight mean reduced transport costs. Unless they run their own delivery transport, however, this saving is seldom a benefit in which brickmakers share. Lighter bricks also make the jobs of building workers marginally easier, provide better thermal insulation and are more resistant to frost damage. On the downside, they are somewhat less strong and durable. If density falls below 1,400 kg/m^3, bricks would likely not be durable enough for use in construction.

When matured at the right moisture content, clay soils can be readily formed into bricks by throwing, pressing or extrusion as appropriate. When fuels or fluxes are added to the clay the moulding characteristics – mouldability - can change. In such cases, more water is usually added to the mix to regain mouldability. This is likely to mean poorer handling properties and increased drying shrinkage. The alternative is to modify clay moulding or extruding equipment to handle drier mixes, which is easier and more cheaply said than done. Highly plastic or sticky clay can be subject to significant shrinkage when dried and fired. This can lead to distortions and serious cracks in bricks that render them unusable (Mason, 2000a). The traditional remedy is to add sand to the clay to make it less plastic. Adding a residue or waste, whether as a fuel, flux or bulkerizer, can have the same effect. Some very fine inorganic materials, such as the finest pulverized fly ash (pfa) fraction, can however increase plasticity and cracking.

Even in harsh climate conditions, well made common bricks should last hundreds of years without serious deterioration. (Writing this, I can look out of my office window at a house chimney that does not appear to retain any pointing mortar at all. Thank goodness it's not sitting on my house! Despite being in an exposed coastal location, however, the bricks in the chimney look almost as good as new, retaining smooth faces and sharp corners.) Many artisanal brickmakers underfire their bricks. Usually, they do this either because they cannot get enough fuel or because production is poorly managed. At best, the resulting bricks are likely to remain in good condition for decades rather than centuries. Abrasion, wind, rain, mortar and render movement and, in some locations, frost, attack and erode the bricks. Adding wastes as fuels or fluxes can ensure bricks are better burned and hence more durable. Conversely, if the bricks are significantly more porous as a result of the addition of waste, durability may be reduced.

Efflorescence is manifest as unsightly white deposits or stains on the surface of brickwork. Though it is not aesthetically pleasing, efflorescence is not normally damaging and is not an indication of poor brick quality. There are actually three categories of efflorescence: lime bloom, lime weeping and crystallization of soluble salts. Lime bloom usually disappears in the longer term due to weathering. It may occur either because bricks have been made from clay with a significant lime content or when lime has been introduced to the process in a fuel or other additive. Well distributed in powder form, lime can act as a flux in brick firing. If it is present as larger stones, however, it will heat up, change form, expand, and is likely to burst out of bricks, thereby ruining their appearance and sale value. Lime weeping is generally seen at cracks and joints in older brickwork and becomes a permanent feature. Crystallization of soluble salts usually takes place where bricks have been produced with water that has a high sodium chloride content. In the extreme this form of efflorescence can cause minor damage, such as swelling and cracks, in brickwork. In some cases using wastes in brickmaking may cause one or other of these types of efflorescence. Other wastes could also cause changes in the colour or hue of bricks. Consumers may well prefer a brick of the appearance they are familiar with. They may even associate a certain hue, the proverbial brick red for instance, with strength or durability. So, regardless of the actual physical properties, brickmakers may have problems selling bricks of a changed hue.

Reduction occurs when insufficient air gets to the fuel. The result is inefficient burning of bricks and fuel that remains partly unburned. If fuel has been incorporated into the brick, the fuel at the core may be black and only part-burned. Also, when bricks are placed too close together in the kiln, black reduction spots can occur where their surfaces are in contact. Bricks with reduction cores and spots are more likely to exhibit substandard properties. Moreover, customers are likely to decide that reduction spots spoil the appearance of bricks and hence be reluctant to buy them. In some cases, problems due to reduction can be solved by modifying the kiln to promote air flow. Most simply, this may be achievable by increasing the spacing of bricks. Forced draught, employing a fan, would be a more extreme and expensive solution at the other end of the spectrum.

Fuel choice and environment

The energy requirement of brickmaking varies widely and depends on many factors, particularly the type of clay used, as well as the drying and firing process employed. Some clays naturally contain organic matter. This acts as fuel when bricks made of that clay are fired, reducing the exogenous energy requirement. The presence of too much organic matter results in a friable and inferior brick, however. Specific energy consumption can range widely from 1 to 12 megajoules per kilogram of fired brick (Russell, 1996). The lower figure could conceivably be achieved in a particularly efficient kiln, such as a continuously operated Vertical Shaft Brick Kiln (VSBK). The higher figure could correspond to an artisanal brick clamp of a few thousand bricks that is fired with insufficient wood and produces under-burned bricks.

Depending on the characteristics of clay or clay-mix used, the temperature required for sufficient vitrification is in the range of 900 to 1,300°C, though it is quite common for the kilns of small-scale brickmakers not to reach such temperatures. Such a relatively high temperature specification indicates that quite a high grade of fuel is required. If wastes and residues are to wholly replace the conventional fuels used in brickmaking, they would evidently need to be of a similar grade, i.e. have a similar calorific value per unit mass and exhibit comparable combustion properties. A significant proportion of the primary fuel can be replaced with a lower grade of fuel in the form of waste, however, albeit in greater quantity. Brickmakers sometimes use inferior fuel because conventional fuel is too expensive or unavailable. Very inferior fuels include rags soaked in used engine oil, plastic wastes, old tyres, and small diameter green wood stripped from bushes and shrubs. The use of such fuels is likely to be an important reason why the bricks produced by some brickmakers are of poor quality and also why brickmakers in some areas are considered a polluting nuisance.

Apart from the levels of pollution produced when burned, the characteristics of a fuel that affect its suitability for brickmaking are its physical properties, calorific value, volatile and ash content. Critical physical properties include particle or lump size, porosity, and structural integrity in the kiln, the latter determining whether or not it crushes under pressure when placed in layers between bricks, for example. Another consideration, when the fuel is added to the clay in the brick, is water absorption. If this is high, the water needed to mould the clay is increased. When using relatively fine fuels, such as sawdust, or coffee or rice husks, a simple method is to distribute the fuel between the layers of bricks, though running the risk of reduction. Fine fuels are patently not very suitable for burning in fires in tunnels running beneath a kiln. Neither are they readily burned in the grates used with, for instance, down-draught kilns. Quite simple air-blowing equipment can be used to blow fine fuels into remote combustion zones such as these, but technology of this order is relatively complex and too expensive for most artisanal brickmakers to consider.

The calorific values for conventional solid and liquid fuels as well as a variety of agricultural residues and agro-industrial wastes are given in Table 3.1. Assessed on the basis of calorific value alone, a number of wastes appear to have potential for utilization in brickmaking, replacing either all or, more likely, part of the primary conventional fuel. Even though Table 3.1 should be regarded as indicative rather than definitive, agricultural residues such as olive pits, the

Table 3.1 Approximate calorific values of fuels, residues and wastes

Fuel/waste	*Calorific value (kJ/kg)*	*Fuel/waste*	*Calorific value (kJ/kg)*
Olive pits	21,400	Plastics	37,000
Olive residues	1,260–18,680	Sawdust	15,900–18,000
Rice husks	12,100–16,000	Wool wash water	
Rice husk ash	2,300	treatment sludge	1,750
Rice straw	15,100	Waste down (textile	18,900–29,400
Vegetable matter	6,700	industry)	
Maize stalks	17,500	Exhausted mineral	7,140
Maize cobs	14, 000–18,200	oils	
Coconut shells	20,100	Waste engine oil	25,000
Groundnut shells	20,100–21,500	Coal-mining wastes	3,530–5,800
Walnut shells	21,100	Petroleum coke	1,470–33,180
Coconut pith	12,850	Fly ash	2,100–11,640
Bagasse-5.25	18,880	Rags	16,000
Coffee husks	16,600	Dust and cinders	9,600
Wood (15% moisture)	15,000	Paper industry sludge	7,000–19,000
Charcoal (2% moisture)	33,000	Paper	14,600
Commercial butane	58,000	Coal	23,000–29,000
Sewage sludge	10,000–23,000	Diesel fuel	44,000
		Heavy fuel oil	42,000

Sources: Dondi, Marsigli and Fabbri, 1997; Lardinois and Van de Klundert, 1993; Haleja et al., 1985; Mason, 2001.

shells of a variety of nuts and the residues from maize harvesting seem to hold particular promise as fuels. The same can be said of some industrial wastes, including plastics and engine oil (although note the different estimate given for 'Exhausted mineral oils'), along with the agro-industrial residue sawdust. A significant drawback with some wastes is evidently their widely variable calorific value, for example sewage sludge, paper industry sludge, and waste down from the textile industry.

Volatiles are substances driven off as gases or vapours when materials are heated from ambient temperatures to a few hundred degrees Celsius. If fuel is fed directly into the kiln, volatiles burn contributing their fuel energy to the process. Volatiles can present a problem when fuel is included in the brick body or distributed throughout the kiln. In these cases the transfer of heat through the kiln precedes ignition, i.e. fuel may be warmed to a few hundred degrees before it actually catches fire. Hence, some volatiles are driven off without igniting. Their calorific value is therefore lost from the process. Moreover, the release of unburned volatiles increases air pollution. Volatiles from wastes can include substances such as benzene or aldehydes and prolonged exposure to high concentrations of such substances can be harmful to health. With wastes that have been used in brickmaking, however, high concentrations and prolonged exposures do not usually result.

Considering air pollution, some wastes such as pfa can actually reduce the emission of particulates or soot. In comparison to conventional fuels, if wastes that are likely to be used in brickmaking are burned effectively, then they do not generally have a significant impact on air pollution, i.e. there is little or no change. Moreover, when we consider wastes that are otherwise disposed of by burning, there is a beneficial impact if they are used as fuel substitutes in brickmaking; the pollution that would have resulted from the combustion of the primary fuel is saved. A problem with wastes is that they may burn at markedly different temperatures from conventional fuels, which may remain the primary fuels in the brick firing process. If either the primary fuel or the waste experiences incomplete combustion to a significant degree, the result will be an increase in smoke.

Cinder 'ash', from boilers, domestic hearths and cookers, is often predominantly carbon and so potentially still combustible in brick kilns. The ash content of these 'ashes' is the non-combustible fraction, predominantly silica (silicon dioxide), alumina (aluminium dioxide) and, in some cases, calcium oxide. Non-combustible ash is a concern with wastes in general. If the waste is incorporated into the brick, such material is most likely to contribute to ceramic bonding and does not represent a problem. In the case of calcium oxide there may be a minor problem with efflorescence in the form of lime bloom, however. If wastes are introduced into the kiln in other ways, there may be a problem with non-combustible ash building up and obstructing the flow of combustion air. This can result in increased levels of incomplete combustion overall and so increase air pollution.

Values for the non-combustible ash content by weight of some agricultural wastes are given by Lardinois and Van de Klundert (1993): rice straw 19.2%, rice husks 15.7%, maize stalks 4.9%, groundnut shells 4.4%, olive pits 3.2%, maize cobs 1.7%, walnut shells 1.1%, coconut shells 0.8%. The high proportion of ash remaining after the combustion of rice straw or husks is notable. Due to their high silica content, these by-products produce relatively large quantities of non-combustible ash. Hence, it may be preferable to burn these wastes in a separate firing chamber where the non-combustible ash can be raked out or fall through a grate.

In some cases there is an interesting advantage associated with burning wastes in brickmaking. Fired clay bricks can incorporate the heavy metal compounds found in some wastes. Usually, bricks can 'contain' these harmful substances without significant leaching. Hence, there is no risk to either builders or the occupants of buildings. Brickmaking therefore has potential as a means of disposal of heavy metals, which can otherwise find their way into food chains, for example.

Having considered brickmaking and the use of wastes from the environmental point of view in the previous chapter, let us now consider the environment and fuel choice more from the perspective of brickmakers. We know that wood is the fuel predominantly used by small- and medium-scale fired clay brick- and tile-makers in developing countries. Charcoal, derived from burning wood, is also employed by the sector in certain locations, usually where the fuel has had to be brought in over some distance. Where it is readily available, for example in the northern states in India, coal can be the fuel of choice. Oil is also used in some places. If simple oil burners are used as the only energy input, however, only small clamps or Scotch Kilns can be fired effectively.

For many artisanal brickmakers wood for fuel is becoming increasingly scarce and expensive. In some countries, permits are employed to control the harvesting of timber, especially in designated conservation areas. Although attempts to impose controls are frequently flouted, brickmakers are often unable to guarantee a supply of fuelwood when they want to burn bricks. Either controls are enforced or else corruption means brickmakers have to pay over the odds for illegal supplies of fuel. This makes brickmaking, which too frequently is only marginally profitable anyway, wholly unviable. Significant numbers of brickmakers, especially those producing at the smallest scale, are likely to have other occupations, 'jobbing' as farmers or petty traders, as common instances. They may only produce bricks seasonally. Losing income from brickmaking can critically affect the situation of many brickmakers and their families, however. From being reasonably secure, perhaps even able improve their lot, the loss exposes them to the risk of extreme poverty.

If brickmaking operations are to be kept relatively simple and low-cost, there are technical constraints on using oil as fuel. Firstly, it is quite difficult to ensure that bricks are evenly burned throughout the kiln when using only a simple oil burner. Kilns have to be relatively small so that the heat produced can reach all the bricks. Unfortunately, though, small kilns are considerably less fuel efficient than larger ones (Mason, 2000b). Moreover, the diesel oil that is usually the only available fuel for oil burners is not well suited to firing bricks at the high temperatures required. Technically, a better choice would be a specially formulated oil, such as the Bunker C type. This will probably be difficult or impossible for brickmakers to obtain, however, especially in rural areas. A general consideration is that price of oil globally is subject to a continual increase. In simple terms, this is largely because demand is growing faster than supply. Indeed, the price may be set to increase more steeply and erratically in future (Campbell, 2005). Given the price, supply difficulty and the possibility of an impending crisis, then, it is surely not a god idea for brickmakers to become reliant on oil-based products as their fuel (Deffeyes, 2005). Indeed, as is widely acknowledged, those of us already oil-dependent will have to make some tough decisions in the not too distant future (Roberts, 2004).

With regard to firing bricks with coal, small-scale brickmakers generally use the lowest grades available because these are the cheapest. Unfortunately, they also tend to be the most smoky and polluting. Typically, brickmakers operate in clusters, occupying a site where the soil is suitable to their purpose and that is close to markets in a town or perhaps a network of villages. Local residents, especially the wealthier ones, are often opposed to brickmaking activities where coal is burned because of the pollution. Small-scale brickmaking can also be an obvious and easy target for administrators from government agencies keen to demonstrate that they are taking action on pollution: brickworks are hard to hide and are likely to be located within a convenient distance of the office. Unless they can demonstrate that they are taking active steps to change to less polluting fuels, the activities and hence livelihoods of brickmakers are often placed in jeopardy. While the wealthy and the administrators are likely to be driving around in gas-guzzling and carbon-emitting vehicles made with non-recyclable materials and technologies, then, brickmakers can be penalized for local pollution: regulation seldom adds up to a rational system, particularly with respect to complex environmental issues.

Small- and medium-scale brickmakers in developing countries may not be considered as significant consumers of energy when compared with, for example, users of wood stoves in the household sector. Nor, by similar token, are such brickmakers responsible for a massive share of pollution. Because they are often clustered in a relatively small area in groups of tens or hundreds, however, kilns and brickyards are visible and odoriferous from some distance away. Thus, although brickmaking may have a minimal impact on other elements of the environment, it is still likely to be judged a highly polluting sector based on very evident air pollution. Furthermore, where brickmakers predominantly use local wood for fuel they contribute to an obvious loss of tree cover. Apart from the environmental impact of this, including degradation of the aesthetics of landscape, brickmakers may be competing with local households and other small-scale enterprises such as bakeries and restauraunts for increasingly scare wood. This can cause resentment and ultimately conflict.

Environmental legislation is having an increasing influence on the lives of some brickmakers. Such legislation is likely to be implemented from the perspective of the conservation of resources and air pollution, ignoring socio-economic and cultural aspects of the environment. In some cases, particularly in more remote rural areas, brickmakers may escape the threat to their livelihoods that such conservation legislation poses. In India, on the other hand, brickmakers tend to operate in medium or large clusters. As we have discussed, this makes their environmental impact very apparent. In such a densely populated and bureaucratic country their chances of escaping the attention of environmental agencies are very slim. As a consequence, the operation of the traditional Bull's Trench Kiln, which usually burns coal, is increasingly restricted throughout India (Raut, 2002). In neighbouring countries such as Nepal, where legislation on pollution is not so rigorously implemented, this type of kiln is still used extensively, however. In Bangladesh this is virtually the only type of kiln used to produce bricks; though the government has imposed, through environmental legislation, an increase in chimney heights to 10.5 metres enabling better combustion and reduction of polluting gases.

The environmental legislation that affects the small-scale brickmaking sector, then, tends to focus on air pollution and the conservation of resources, particularly indigenous woodlands. The former aspect prohibits brickmakers from burning fuels such as low-grade coal, used rubber tyres or waste engine oil. The unfortunate impact of implementing such well-meaning and seemingly beneficial legislation is that brickmakers are forced out of business. In Tanzania, for example, there are now significantly fewer artisanal brickmakers than in 1990 due to restrictions on the cutting of trees in remaining forests (Merschmeyer, 2003). Transport costs render alternative fuels too expensive. Little has been done to mitigate the effects that legislation has had on the livelihoods of brickmakers. There have been no government schemes to set up sustainably managed fuelwood plantations, for example. Neither have production technologies using alternative fuels been officially promoted. The limited interventions that have been made have involved NGO initiatives.

By contrast, in Ciudad Juárez in Mexico significant success has been achieved without application of restrictive laws and regulations (Blackman, 2000). Brickmakers have moved away from using very polluting fuels, such as used tyres and old engine oil. One critical factor in achieving this change has been rising awareness, particularly about pollution and health. Moreover, brickmakers and surrounding communities have been thoroughly involved in developing fuel use and monitoring procedures. Although these procedures were not enforced through any conventional legal channels and largely relied on the voluntary cooperation of brickmakers, there were relatively few defaulters. The censure of the community and other brickmakers is a powerful inducement not to go back to using polluting fuels.

The potential for using wastes

Many types of domestic, industrial and agricultural wastes have a calorific value and so have potential for use in domestic or industrial processes that require heat. Some wastes, for example pfa, coal-ash cinders and rice husks, can be cleaner-burning and so less polluting than conventional fuels such as coal or wood. Certain industrial wastes can act as fluxes, meanwhile, lowering the firing temperature and thence the energy required to form the ceramic bond in clay mixes. This will obviously reduce fuel use and the associated cost proportionally. In the case of the bricks typically produced by small-scale enterprises in majority world countries, ceramic bonds are seldom wholly formed. What is achieved is sufficient vitrification to invest the brick with the necessary physical properties to perform its function. That said, the same energy-saving argument applies vis-à-vis fluxes. Other types of waste can act as grog or opening agents. When added to the mix for moulding bricks, grogs 'invade' the dense microstructure of clay. The resultant material matrix helps to promote

Photo 3.2 Close-up view of a brick. Credit: Practical Action/Zul.

vitrification throughout bricks when fired. The addition of grog also reduces firing cracks and other potential problems such as black cores.

Depending on the nature of the waste added, the physical properties of burned bricks may be either improved or impaired. Although some wastes act as fuels, fluxes or grogs, they may also have a negative effect on brick quality. Some metallurgical sludges, in particular, cause bricks to achieve lower strength and resistance to weathering. Furthermore, drying and firing shrinkage may be increased along with efflorescence. Sometimes the addition of a waste improves one property at the expense of another. Adding sawdust to the clay mix, for example, may reduce the density and hence mass of the final brick, which is beneficial from the point of view of transport and perhaps use. Mechanical strength tends to be reduced by the addition of sawdust, however. And if mechanical strength falls below either regulatory or practical limits, then customers will reject bricks. Striking the balance, i.e. knowing how much waste can be beneficially added, is therefore essential.

There are many wastes with the potential for utilization in brickmaking. These include sawdust from sawmills, rice husks and stalks, coffee husks, coconut shells, maize stalks and corncob cores, bagasse fibres from sugar making, olive pits, groundnut shells, sewage sludge, textile wastes and washes, petroleum coke, tannery wastes and sludges, papermaking wastes and sludges, blast furnace slags from steelmaking, colliery mining wastes, boiler and power station cinders and ash, and pfa. Taken together, huge quantities of these wastes are generated globally. Data to assess the exact quantities available are hard to come by. Safe to say, though, that these wastes could theoretically provide enough energy to supply all the world's brickmaking operations many times over.

To illustrate the point, it is worth giving a few examples of the scale of waste generation. Let us consider agricultural wastes initially. Globally, maize production, including the mass of the cob, is estimated to exceed 600 million tonnes per annum. The figure for rough rice, which includes the husk, is similar. Around the world, processed coffee production is estimated to exceed 7 million tonnes annually. The mass of husks left over would be more than double this figure, perhaps 15 million tonnes. Worldwide, about 10 million tonnes of olives are harvested annually, some 50 million tonnes of coconuts, and around 30 million tonnes of groundnuts. As a guide, roughly half the weight this groundnut yield is made up of the shell, a highly combustible waste. Every year the world produces around 150 million tonnes of sugar from sugar cane. An equivalent weight of residue, combustible bagasse, is generally discarded and left to rot.

The scale of generation of agro-industrial and industrial wastes and residues is also vast. In Asia, for example, about 50 million tonnes of sawn timber is produced. Though it is relatively small, actually a vast quantity of sawdust is discarded as a by-product. India is one of the world's leading coal-mining nations, extracting and processing an estimated 130 million tonnes of coal per annum. A significant proportion of this coal is burnt in electricity-generating power stations or industrial boilers, many of which are not very efficient. The quantities of pfa and cinder produced as waste are therefore considerable, and both will likely retain significant calorific value,

In most majority world countries only a small proportion of the combustible waste generated from agricultural production is used for fuel in domestic or industrial applications. Somewhat amazingly, to the energy-conscious engineer at least, such wastes are neither used directly nor processed into fuel briquettes, nor pyrolized into charcoal, nor utilized for the production of biogas or producer gas. These wastes are, then, wasted. As such they typically represent a management problem that is 'solved' by dumping or burning. Dumping can result in very negative environmental impacts, such as surface and groundwater pollution. Disposing of wastes by burning them in the open air, meanwhile, causes air pollution and greenhouse gas emissions. It may also give rise to a further problem, namely disposing of the ash.

Often burning residues for disposal purposes is undertaken in the same areas where people experience problems acquiring conventional fuels such as wood, coal, charcoal, diesel oil or kerosene for their homes or businesses. Brickmaking is just one of a number of small-scale industries that could use much greater quantities of agricultural and industrial residues as fuel. Brickmakers could also utilize some residues to modify brick or tile properties. There is evidently a compelling argument for assessing the environmental impact of using wastes in brickmaking against methods of disposal such as dumping and burning. As opposed to disposal of wastes by burning, surely using them as fuels in brickmaking is a win–win scenario? Unless the combustion process is extremely inefficient and smoky, burning wastes productively will not cause additional pollution or greenhouse gas emissions, but will help conserve the resources of the fuel that wastes replace. For brickmakers, moreover, problems of fuel scarcity and affordability could potentially be overcome at a stroke.

Our review so far has indicated that wastes constitute a veritable gift to brickmakers. Almost inevitably, the scenario is not wholly positive. There are a number of problems that would need to be overcome to ensure that the use of wastes could be widely adopted. Firstly, it is not often that brickworks are conveniently sited near producers of agricultural or industrial waste. Sites with clay soils suitable and available for brickmaking may be in locations distant from where soils are suitable for mass agricultural production, for example. Some wastes, such as rice, groundnut and coconut husks, straw and bagasse, are quite bulky. Hence, the necessity of transporting them even a few kilometres can make their utilization financially unattractive.

Although the global quantity of agricultural wastes is vast, their production density tends to be low and a potential collector and distributor might have to visit a goodly number of small farms to make up a decent load. In addition, their availability is bound to be seasonal. The best year-round sources of agricultural wastes would be sizeable food-processing factories such as rice mills, olive oil, groundnut oil or sugar-processing plants. Once again, it is unusual to find such facilities situated in close proximity to brickmaking ventures. The situation is most often similar for industrial wastes such as pfa from coal-burning power stations and even sawdust from sawmills.

Apart from problems of location, there are technical problems associated with burning wastes. For one thing, the soils used in brickmaking vary. Though they are generally termed 'clays', actually the clay content varies considerably, as do other mineralogical constituents. Certain clays are therefore better suited to the incorporation of certain wastes. A high-clay soil, for instance, will generally benefit from the addition of a greater percentage of sawdust than will a more sandy soil. Adding sawdust to bricks made from sandier soils tends to make them porous and weak.

With certain wastes it is only possible to replace part of the fuel without adversely affecting the properties of the bricks. This is typically the case with adding rice husks to a clay mix, for instance. Rice husks have a low energy density and incorporating more than a few per cent by mass results in bricks that have many voids and therefore absorb water too readily and are mechanically weakened. In such an instance, a significant proportion of another type of fuel would still be needed. This complicates the production process. How much of each fuel is required? Can fuels be supplied consecutively? How are the two fuels to be distributed and burned? If the waste fuel is ground or pulverized, we have indicated that generally a greater proportion can be utilized as fuel incorporated in bricks. There are no free lunches, however, and such processing adds to costs and has an associated environmental impact.

Other process changes may be necessary for brickmakers to take advantage of wastes as fuel. When the fuel is incorporated into the body of the brick, for example, mixing and moulding techniques may need to change. When wastes are burned separately, methods of charging the kiln with bricks and fuels will need to be modified. It may be critical in the acceptance of new technology that process changes are minimal. If the changes required are relatively small, inexpensive and not too technically demanding, then understandably brickmakers will be much more likely to adopt the technology. If major change is demanded, however, for example if brickmakers need to change from traditional slop-moulding to extrusion, then this is likely to be a very considerable deterrent.

Though some wastes burn well, others contain a high proportion of volatile materials. As we have discussed, if such wastes are burned in certain types of brickmaking kilns, the volatiles are actually driven off at a few hundred degrees Celsius without catching fire. The calorific value of the volatiles is therefore lost and their emission from the kiln adds to the pollution.

The calorific values of wastes are often lower than those of conventional fuels, i.e. the energy density is less. We know that this has to be taken into account when considering using wastes as fuels. Can the necessarily greater mass of fuel be successfully introduced into the firing process? More difficult to deal with than lower calorific values per unit mass is variability. Some wastes, especially those emanating from industrial sources, can be quite variable in their chemical or mineralogical characteristics and thence calorific value and other key brickmaking properties. In such instances, brickmakers would have to continuously monitor changes in the properties of the waste and adjust proportions and burning techniques to take account of variations. This significantly adds to the complexity of the process as well as costs. In general, small-scale brickworks are not equipped with either the capital or the skills to run an on-site laboratory.

The utilization of wastes as fuels or additives in brickmaking is still far from widespread. Only sawdust and boiler cinders or ashes are used to any great extent in commercial brickmaking. Other wastes have perhaps been the subject of laboratory-scale trials or tested in pilot projects in the field, usually over short periods. It should be noted that most such trials with wastes and residues have been undertaken by industrial-scale companies in the West rather than by artisanal brickmakers in the majority world. Although in some cases the trials yielded promising results, in the main they were discontinued. In general, brickmakers in the West have had no trouble sourcing conventional fuels - with which they are more familiar - cheaply enough. To date, there has been no great incentive to look for alternatives, certainly not those that would require technology or process changes to introduce. The results of most trials have only been published in Western technical journals or internal reports, which are not readily available or accessible to small-scale brickmakers in the majority world. So, information on the use of wastes in brickmaking has so far not been widely or appropriately disseminated.

Some concluding remarks on ways of working

So, can promoting alternative fuels such as wastes or residues improve the viability of small-scale brickmaking operations? Can it reduce their environmental impact? Are these two possible outcomes even compatible? Consider that other innovations in brickmaking can have beneficial affects on livelihoods and the environment, for example:

- better training of brickmakers;
- better clay preparation and moulding of bricks;
- introducing better quality control;
- building larger, more efficient brick clamps;
- replacing clamp firing with kilns, which reduces heat losses;
- introducing more efficient kilns, such as the VSBK.

Small-scale artisanal brickmakers, almost by definition, do not have huge surpluses to invest. In contrast to introducing wastes, kiln modifications or changing to a different type of kiln can be very expensive. Indeed, such measures may not be realistic options at all. In Ciudad Juárez, Mexico, although they have consistently expressed interest, none of the brickmakers have installed the more fuel-efficient and cleaner burning kilns that have been developed specifically to suit their needs (Blackman, 2000). The relatively high cost of the improved kilns and the difficulties most of the brickmakers would face in raising capital have proved to be insurmountable obstacles.

This cautionary note on capital investment made, the other innovations listed are probably best implemented as a programme. Training, improved production practices, quality control and the utilization of waste can, and preferably should, all go hand in hand. A specific point in favour of interventions in fuel use practice is that between 25 and 50 per cent of the production costs of small-scale brickmakers are fuel costs. Often their fuel costs are higher than their labour costs. Any potential for saving fuel costs is therefore likely to strongly appeal to brickmakers. As we have recorded, this is especially the case if there is not too great a capital outlay.

Brickmaking is a skilled and intensely practical occupation that incorporates perhaps generations of experience. Part of the challenge of promoting the use of wastes in brickmaking is, then, to do with overcoming a natural conservatism vested in tradition. New theoretical information does not become embodied knowledge overnight. For NGOs and others working in the sector, choosing appropriate partners amongst brickmakers is critical when seeking to disseminate a technology. The NGO will wish to form trusting relationships with brickmakers who have the respect of their community because they have the community interest at heart. Moreover, when these brickmakers enter into research and development partnerships, they must be sheltered from financial and any other associated risks. Technology change is a learning process that necessitates training and may, initially at least, mean loss of production. Almost inevitably, therefore, it will mean reduced profit in the short term. The price of technology change cannot be at the expense of the vulnerable livelihoods of small-scale brickmakers.

As we have stressed throughout, many small-scale brickmakers in the majority world operate on the margins. They have to overcome numerous difficulties, some of which threaten to close their operations altogether. When these brickmakers undertake innovation, they necessarily want and tend to do so cautiously, making small and incremental changes. They cannot afford the risk of making quantum leaps in practice, such as switching from a tried and tested fuel like wood. If they cannot make small changes work both technically and financially in a short time, moreover, they do not have the resources to persevere. If they can visit a successful pilot demonstration project or gain insights from a brickmaker in their area, however, and if they can see that something works in practice in similar circumstances to their own, they are much more likely to take an interest in adopting an innovation. These brief observations on integrated technology development and participative ways of working made (for details see Mason, 2001, and the references used therein), let us move on to consider our case studies.

CHAPTER 4

Trials with coal-dust, coal-dust briquettes, waste oil, rice husks and sawdust in Peru

Emilio Mayorga Navarro and Saul Ramirez

History, culture and politics

The cultural and political history of Peru is ancient, fascinating and extensive. It can only really be touched upon in this introduction to the nation, a quickfire guide to 10,000 years of a land and its peoples. From 2500 BCE there are records of a succession of cultures in the territory we know as Peru, peoples who initially engaged in subsistence agriculture and livestock breeding. The highly developed Chavín de Huantar culture flourished between 900 and 200 BCE. During the 4th century BCE, the Chimu and Nazca cultures developed markedly. Their textiles, metallurgy and highly technical irrigation systems are acknowledged as outstanding. By the 6th century CE it was the Tiahuanaco culture that was in the ascendancy. The 12th century CE marked the advent of the Incas, currently the best known of the indigenous cultural epochs. The Inca empire centred on their capital, Cusco, 'the Navel of the World', and they expanded their territory greatly within a very short time. They were hydraulics experts and excellent farmers who cultivated their land with a system of benched terraces.

The Spanish conquest began in 1531, when the Inca empire had been weakened by a war of succession between the armies of the brothers Huascar and Atahualpa. The Spaniards arrived in Cusco in 1533 and Atahualpa, the last 'Sapa Inca' or ruler of the Inca empire, was killed on the orders of infamous conquistador Francisco Pizzaro. Two years later the Spaniards founded Lima, Peru's capital city to this day. Ensuing rivalries between the conquerors resulted in successive civil wars, which continued until 1547. The Peruvian Viceroyship was then consolidated, comprising most of the conquered territories of South America. By the time Spanish domination ended, political machinations had reduced the Viceroyship to the current territories of Peru and parts of Bolivia, Brazil, Chile, Colombia and Ecuador.

After the repression of the uprisings of Santos Atahualpa in 1742 and Tupac Amaru from 1780 to 1781, the final struggle for independence from the colonial power could not begin until 1810. Jose de San Martin, backed by the armed forces of Argentina and Chile, was finally able to proclaim independence on 28 July 1821. With the help of Colombian, Argentinean and Chilean forces, under the direction of Simon Bolivar, 'El Libertador', this independence was secured

Figure 4.1 Map of Peru

in 1824 in the battles of Junin and Ayacucho. A troublesome period followed, however, characterized by anarchy, civil wars and disputes with neighbouring countries. The worst of these disputes resulted in Chile declaring war on Peru in 1879. Peru did not recover from this extremely bloody war, and subsequent civil conflict, until perhaps the fourth decade of the following century.

The architect Fernando Belaunde Terry was elected president in 1963. In 1968 he was ousted by a nationalist military coup headed by General Juan

Velasco Alvarado, who governed the country until 1975. The general introduced a number of socialist measures, the most prominent being the nationalization of the oil and fishing industries, and reforms of the education and agricultural systems. A serious economic crisis gave rise to another military coup in August 1975, this time led by General Francisco Morales Bermudez. Bermudez governed the country until 1980 when he handed power back to the democratically elected president, Fernando Belaunde Terry. Unfortunately, President Terry was unable to deal effectively with the economic crisis or defeat the bands of armed insurgents who had emerged to inhabit the political landscape.

In 1985 Alan Garcia Perez came to power. This was the first time in Peru's 60-year history that a representative of the populist APRA (American Popular Revolutionary Alliance) had been elected president. It was to be an unpopular disappointment. During the government of Perez the economic crisis was further aggravated and the activities of the insurgents increased. Perez's unwillingness to service the nation's foreign debt, which amounted to a massive US$ 14 billion, resulted in his period in office being characterized by international economic and financial ostracization. Alberto Fujimori was elected president on 28 July 1990. His government was characterized by the reinstatement of Peru into the International Monetary Fund, a general improvement in the population's standard of living, an increase in production and exports, the resolution of long-standing border disputes with Ecuador and Chile, and a more effective struggle against the insurgents. Abimael Guzmán, the leader of Sendero Luminoso (Shining Path), a Maoist guerrilla organization, was captured in 1992, and since then insurgent activity in Peru has been only minor and sporadic. Unsurprisingly, given this list of successes, Fujimori was re-elected in 1995 and again in 2000. Due to problems related to corruption and political instability, however, he resigned on 21 November 2000. Dr Valentin Paniagua was declared provisional president and remained in office until July 2001, when Dr Alejandro Toledo took over as constitutional president of Peru until elections scheduled for 2006.

Among its main achievements, the new government has maintained a positive macroeconomic environment and started the regionalization process, whereby power will gradually be transferred to the regions. Notably, in the context of this book particularly, the government has also implemented a dynamic low-cost housing programme in urban areas. This programme is complemented by a credit scheme for potential buyers, and measures to regularize land titles. The effect of these combined measure has been an increase in housing investment and hence presumably an increased demand for bricks and building materials. Given persistent evidence of corruption at the highest levels, though, the current political situation is unstable. As a product mainly of the regionalization process, moreover, laws are constantly being amended and on occasion contradictory laws have been enacted and enforced. In most recent times, this crisis in the judiciary does seem to have improved substantially, however.

Box 4.1 Key comparators for the United Kingdom

Land area: 241,590 km^2

Population: 60,441,457 (July 2005 est.)

Age structure: 0-14 years 17.7%, 15-64 years 66.5%, 65 years and over 15.8%

Population growth rate: 0.28% (2005 est.)

Infant mortality total: 5.16 deaths/1,000 live births

Life expectancy: men 76, women 81 years

Unemployment rate: 4.8%

Literacy: 99%

Population below the poverty line: 17%

GDP/capita (purchasing power parity): $29,600 (2004 est.)

GDP real growth rate: 3.2% (2004 est.)

Exports: manufactured goods, fuels, chemicals, food, beverages, tobacco

Imports: manufactured goods, machinery, fuels, foodstuffs.

Climate, land and people

Peru covers a total area of some 1,285,217 km^2. Given the geographical location it should be a uniformly tropical land, with abundant rainfall, high temperatures and lush vegetation. Instead it is a country with immense climatic diversity. The combination of the Andes mountain range, the Humboldt ocean current and El Niño means Peru experiences almost every type of weather imaginable, resulting in diverse ecological conditions across the nation. In our particular sphere of interest this means that Peruvian brickmakers deal with a range of soil types and landscapes. They also have diverse range of agricultural wastes available in the various regions.

The northern coast, between Tumbes and Piura, has a very hot, semi-tropical climate with an annual average temperature of 24°C, regular rainfall between January and March, and high humidity. South of these departments, between Lambayeque and Tacna, the climate on the coast becomes subtropical with temperatures of between 18 and 21°C and humidity between 90 and 98 per cent. In the highlands it is much colder, the temperature ranging from just 6 to 16°C. At altitudes over 4,500 metres above sea level there are snow-capped mountains and glaciers and in the Andean plateau of the Puno department the weather is particularly cold. Between 2,500 and 3,800 metres, the rainy season occurs between December and April. Above 3,800 metres this rain turns to snow and hailstones. In the Amazon region, meanwhile, it is hot and humid with abundant rainfall all year round. Between January and April this rainfall is particularly high, providing the conditions for river navigation. The area with the heaviest rainfall is the Lower Jungle area or Amazon Plain. Average annual temperatures in this region vary between 16 and 35°C. Somewhat confusingly, but following flawless logic, the lowest temperatures are recorded in the Higher Jungle and the highest in the Lower Jungle.

The three main regions of Peru are, then, the coast, the highlands and the Amazon region. The coast covers 10% of national territorial area and is a narrow strip between the Pacific Ocean and the Andes mountain range. The majority of the people on the coast are of mixed race, predominantly Hispanic and Indian. The highlands cover 30% of the territory, comprising a strip that runs parallel to the coast and which is dominated by the Andes mountain range. The people here are predominantly indigenous, mainly of Quechua and Aymara origin in the highland plateaus of the Puno department. The Amazon region covers the area east of the Andes, equivalent to 60% of the territory. Its urban population are of mixed race, but the people in rural areas are predominantly Amazonian natives from a variety of tribes. The country has an inverse population distribution, i.e. the largest population, equivalent to 50%, live on the coast, 40% in the highlands, and only 10% in the expansive Amazon region. In all regions, the population is increasingly concentrated in cities and towns, putting pressure on housing and service infrastructures.

On the coast, the majority of the population inevitably live in cities and are involved in trade or industry. These are also the dominant economic activities in the large towns of the highland and Amazon regions. In rural areas of the coast, people are mainly employed in arable farming, growing cotton, sugar cane or rice, or in livestock activities, particularly fishing and poultry production. In rural areas of the highlands, the main economic activity is mining, but here is also farming, with corn, potatoes and sweet potatoes being the main crops. Cattle farming and the rearing of llamas are quite common occupations in Cajamarca in the north and in Arequipa in the south. In rural areas of the Amazon region, economic activity is centred on fishing, logging, and harvesting cassava and local varieties of fruit. Native communities remain essentially hunters and gatherers in a subsistence economy.

Table 4.1 Contribution to GDP and employment by sector

Economic sector	*%age of GDP*	*Number of jobs*
Farming & livestock	7.60	1,906,305
Fishing	0.72	47,366
Mining & hydrocarbons	4.67	69,413
Manufacturing	15.98	754,165
Electricity & water	1.90	17,523
Construction	5.58	225,029
Trade	14.57	1,145,315
Other services, including government services	39.25	934,351

Note: percentage figures do not add up to 100.

According to estimates for 2005, Peru has a population of 27,946,774. Such estimates are based on the census taken in 1993. A new national population and housing census was taken in 2005, but the results are still pending. Around 32% of the population is under 15, 67% between 15 and 63, and only the remaining 1% are over 64 years of age. The population growth rate is 1.8% a year. It is anticipated that this will continue to decrease, as it has over recent years, to reach 1.3% in 2010. The infant mortality rate is 43 of every thousand children born. Life expectancy, meanwhile, is 72 years for women and 67 years for men. Apart from the gender difference, someone who could expect to live to 72 in an urban area would statistically only live to 65 if they were a rural dweller. Around 73 per cent of the population currently live in urban areas and, following the ubiquitous global trend, this percentage is increasing.

The main social problems in Peru result from the difference in income between the poor mass of the population and the rich elite. Government corruption is also a source of concern and tension. Furthermore, Peru is the main coca-growing area in the region and drug-trafficking along with a general increase in crime have recently been cited as a prompt for the government to instigate tough measures aimed at ensuring security. The combination of such social problems destabilizes the political situation, and the scourge of insurgent terrorism, which had very largely been overcome, is threatening to return to prominence.

The economy

Peru's quantitative economy mirrors its varied climate and geography. The principal natural resources are marine resources on the coast, mining resources in the highlands and oil and forestry resources in the Amazon region. The national economy depends heavily on exports of minerals and metals. Prices on the world market are subject to extreme fluctuations, however, adding economic insecurity to Peru's political instability. Tourism is a developing sector of the economy but is not without its drawbacks. The ever-increasing number of visitors to Machu Picchu has resulted in severe path erosion and access will now have to be restricted if the site is to be preserved. The advent of such problems will hopefully yield environmental knowledge that can be applied when sites such as the city of Caral are opened to the public. Caral dates back to 2500 BCE and its wonders are sure to attract the avid attention of the tourist industry. Overall, though a lack of adequate infrastructure serves to deter trade and investment, Peru has maintained economic growth in recent years and succeeded in combining that achievement with a stable exchange rate and low inflation.

Though the literacy rate among people over 15 years of age is relatively high at around 88%, job opportunities are extremely limited and unemployment and underemployment are persistent problems. The urban unemployment rate in Peru is 9.4%. Though this figure is relatively low, it conceals a problem of a greater magnitude. More than half of the labour-force are obliged to create their own, usually menial, jobs and so seldom have the opportunity to raise or build-up capital. Hence, Peru's unemployment rate cannot be compared to countries that actually have a much higher percentage of wage-earning workers.

Almost 25% of the population live in abject poverty, with a monthly income of just US$29-37. Estimates of GDP per capita vary quite widely between US$2,922 and US$5,600. Given the difficulty in estimating the contribution of the informal economy, not to mention the illicit economy associated with drug production and trafficking, such disparate figures are perhaps to be expected. GDP is estimated to have risen by 4.1% in 2003, 4.8% in 2004 and around 5.9% by June in 2005. The inflation rate, meanwhile, was 2.5% in 2003, 3.5% in 2004 and 1.5% in the first half of 2005. Peru's most significant imports are petroleum and petroleum products, plastics, machinery, vehicles, iron and steel, wheat, maize, and paper. In addition the nation imports telecommunications and TV equipment, soya products, medicines, and liquid propane. Exports, meanwhile, include copper, gold, zinc and other metals, as well as petroleum and petroleum products. Coffee, fishmeal, and cotton clothing are also significant exports.

Macro-economic problems are the foreign and domestic debts. The foreign debt amounts to US$23,574 billion and the domestic debt is US$4,871 billion. It is estimated that more than 26 per cent of Peru's total budget revenues, around US$14 billion, is earmarked for paying these debts. A further problem is the cost of credit, with high interest rates slowing investment and development. The commercial interest rate for loans in US dollars is 8.6 per cent. The interest rates for consumer and micro-financing purposes are 64 and 54 per cent, respectively. Over the past 15 years, economic policies aimed at attracting investment have not had a significant positive effect. Most foreign investments have been made in the purchase of State companies rather than in new productive enterprises; public goods have simply been sold off at bargain-basement rates to the transnational private sector.

Environmental considerations

The principal environmental focus of Practical Action's brickmaking projects has been deforestation. Forests cover nearly two-thirds of Peruvian territory, which extends to around 128 million hectares in total. The average deforestation rate is 261,158 hectares per year, equivalent to around 715 hectares a day. It is estimated that more than 9.5 million hectares of indigenous forest has already been lost. Around 17 per cent of this deforestation is attributed to logging for self-consumption of fuel, i.e. either the wood is used as firewood, including for brickmaking, or the forest is felled to enable coal mining.

There is national environmental legislation that affects brickmakers but it is ineffective. Law 26258 (December 1993) banned felling in the natural forests located in the departments of Tumbes, Piura, Lambayeque and La Libertad for 15 years. The production, transportation and marketing of firewood and coal from those areas are also banned. Subsequently, Law 27308, or the Forestry and Wildlife Law (July 2000) and Supreme Decree DS 014-2001 (April 2001) regulated the sustainable utilization of forestry resources and fauna. This package of legislation incorporated forestation and reforestation activities into development programmes and authorized the Ministry of Agriculture to impose closed seasons to control the exploitation of native forest and wildlife species. The National Institute of Natural Resources is the authority in charge of establishing the measures to protect species endangered by forestry.

Despite such a sweeping package of legislation, almost all the firewood used for firing bricks is obtained from indiscriminate felling and not from planned and sustainably managed forestry programmes. It is evident that the forestry police in Tumbes, for example, allow loggers to cut down guayacán trees as long as they sign a commitment to plant other trees in their place. This commitment is never enforced. Even at a national level, annual reforestation plans cover only about 100,000 hectares, whereas the annual deforestation rate, as we have mentioned, is more 2.6 times that area. Furthermore, the government's capacity to control deforestation in rural areas is virtually non-existent. In recent years this situation has been still further undermined as the total number of police officers in the country has decreased substantially.

With respect to burning coal, in areas of Practical Action involvement anthracitic coal is used on a large scale only in La Libertad and semi-bituminous coal is only extensively used in Ayacucho. In both areas, coal is marketed so comprehensively that, regardless of the relative environmental impact, it would be difficult to replace it with a financially viable alternative. Practical Action Peru judge that the main problems related to the use of coal-dust are: sulphur dioxide emissions, which can contribute to acid rain; emissions of particles; and nitrogen dioxide, which causes health problems when inhaled. Using the oil burner raises similar environmental concerns. Oil contaminated with chlorine and PCB (polychlorinated biphenyl) produce highly toxic emissions.

Table 4.2 Classification of bricks in Peru*

Type of Brick		*Resistance to minimum compression (kgf/cm²)*
TYPE I:	Low bearing capacity and endurance. Fit for masonry constructions under minimum service conditions.	600
TYPE II:	Low bearing capacity and endurance. Fit for masonry constructions under moderate service conditions.	700
TYPE III:	Average bearing capacity and endurance. Fit for masonry constructions for general use.	950
TYPE IV:	High bearing capacity and endurance. Fit for masonry constructions under rigorous service conditions.	1,300
TYPE V:	Very high bearing capacity and endurance. Fit for masonry constructions under particularly rigorous service conditions.	1,800

*Peruvian Technical Standard ITINTEC 331.017

The impact of burning waste oil needs to be compared to using it in other ways and to methods of otherwise disposing of it. Apart from deforestation, emissions and pollution, Practical Action Peru noted that important environmental impacts caused by brickmaking are the alteration of the 'geomorphologic and topographic characteristics' of the areas in which clay and sand quarries are located.

In terms of assessing the environmental impact of traditional brickmaking practices, Practical Action are unaware of any survey on a regional scale. A recent study comprised only an initial survey of the impacts of brickmaking enterprises in Cusco and Arequipa, proposing mitigation measures (Piñeiro, 2005). As regards the Practical Action intervention, no specific assessment has been conducted. In the Project Outcomes chapter of *Brick by Brick*, it was noted that, with respect to measuring environmental impact, there was 'more to be done' (Mason, 2001). Unfortunately, funding to continue work in the sector has not been forthcoming and so there is not much progress to report in this regard.

Building materials and shelter

According to the census taken by Peru's National Institute of Statistics in 1993, the average annual growth rate of the housing sector was 2.8 per cent. The census reported that these new houses were mainly improvised dwellings, however, and that they were inadequate for human habitation. Dwellings of this type increased by 14.2 per cent between 1981 and 1993, a rise amounting to 1.5 million houses. During the census, 5.1 million houses were counted. Of this total, 3.5% were improvised dwellings, 3.8% were shacks, 0.6% structures not built for habitation purposes and 0.1% other constructions of an undefined nature. Rural areas are relatively deprived economically and rural housing is therefore especially precarious. Furthermore, population growth and land scarcity compel people, *campesinos*, to invade state and private land in order to build houses for their new families. These houses tend to be rustic structures that lack water and sewage services.

On the national scale there are simply not enough houses. One of the government's main programmes is aimed at meeting the demand in urban areas. The main public institution in charge of the housing programme is the MIVIVIENDA Fund, created for the purpose of promoting access to new housing and to encourage saving to that end. The Fund provides the opportunity for people who would normally be excluded access to mortgages. Major initiatives by the Fund are MIVIVIENDA mortgage loans and the Techo Propio (Own Roof) Programme. MIVIVIENDA mortgage loans offer middle- and low-income people mortgages at less than the market rate of interest. The Prompt Payment Reward initiative makes it financially advantageous to keep up mortgage repayments, moreover. The Techo Propio Programme consists of a direct subsidy (Family Habitation Bond). This Bond is designed to offer lower-income households finance so that they can afford to renovate their homes and attain adequate shelter.

Type III bricks, of good enough quality for general use in building, are really only produced by medium to large-scale plants in or near large cities and are commonly known as industrial bricks. Adding transport to their already relatively high production cost makes it unfeasible to use such bricks in distant rural areas where the purchasing power of the population is very low. It is a similar story with other 'modern' building materials, cement, steel, corrugated iron etc. Most of these materials are manufactured under well-known brand names in areas that can be readily monitored by the appropriate officials and where the consumer has alternative choices. Hence, their quality is largely guaranteed, and this includes Type III bricks

Work with brickmakers

Practical Action's work with brickmakers in Peru was initially born out of concern about deforestation and knock-on environmental and social problems. Deforestation, much of which is caused by illegal logging, is one of the main environmental problems that Peru faces. The indiscriminate felling of carob trees in dry natural forests, mainly on the northern coast, is a significant part of this national problem and many of these trees are being felled as fuel for firing bricks. Practical Action implemented two projects. The first was called 'Energy Efficiency in the Small-scale Brickmaking Process' and was financed by the

Photo 4.1 'King Kong' green bricks drying in La Huaca at the unit of Julio Sánchez, who fires his 7,000 brick kiln with wood, sawdust and rice husks. Photo: Practical Action/ Emilio Mayorga.

UK's Department for International Development (DfID) between April 1996 and March 2000. The objective of this project was to transfer energy-efficient technologies to small-scale brickmakers in 'developing countries' in order to make their enterprises more profitable, particularly by reducing their expenditure on fuels. At the same time, the project aimed to preserve or enhance the environment at every scale: local, regional and global.

The second project was entitled 'Utilization of Rice Husks as a Source of Energy for Brickmakers'. This project was financed by the APGEP-SENREM Programme with USAID funds and was implemented between April 1999 and February 2001. Its main objective was to investigate replacing at least a proportion of the carob tree wood used as fuel in brickmaking with rice husks. The project was located on the northern coast and concentrated on developing techniques for the efficient use of rice husks in the firing process. The main research site was the La Huaca brick 'factory', an extensive site accommodating a number of brickmaking enterprises. La Huaca is a hot and arid, semi-desert location in the Paita province in the department of Piura.

Nationwide, Practical Action Peru's partners in brickmaking projects were small-scale brickmakers who generally operated in the informal sector of the economy. In the majority of cases, brickmakers in a particular enterprise are members of the same family. The size of non-industrial brickmaking enterprises, those we loosely dub small-scale, depends on the geographical area in which they are located and the characteristics of local demand. Apart from the areas where Practical Action's brickmaking projects were located, which are discussed in detail below, there are other very important brickmaking areas in Peru. Of particular note are La Libertad and San Martin, as well as the departments of

Table 4.3 Compulsory requirements for clay bricks: variations in dimensions, distortion, resistance to compression and density*

TYPE	*Variation in size (maximum percentage)*			*Distortion (maximum in mm)*	*Resistance to compression (minimum in kgf/cm²)*	*Density (minimum in g/cm³)*
	Up to 10 cm	Up to 15 cm	More than 15 cm			
I Alternatively	± 8	± 6	± 4	10	Unlimited 600	1.50 Unlimited
II Alternatively	7	± 6	± 4	8	Unlimited 700	1.60 1.55
III	± 5	± 4	± 3	6	950	1.60
IV	± 4	± 3	± 2	4	1,300	1.65
V	± 3	± 2	± 1	2	1,800	1.70

*National Technical Standard ITINTEC 331.018.

Lambayeque, Cusco and Arequipa. Although Practical Action was not directly involved in project work with brickmakers in these areas, the technologies used by brickmakers there were investigated and noted. Indeed, certain of these technologies became the subject of pilot technology transfer initiatives, notably the waste oil burner.

To dispense with the vague label of small-scale, let us summarize the characteristics of the brickmaking enterprises involved as partners in the projects implemented by Practical Action. In Pampa Grande and El Edén in Tumbes, brickmakers work with small kilns of between 6,000 and 11,000 bricks. An average kiln would consist of 8,000 bricks and measure approximately 3.33 m x 2.90 m x 3.40 m. A typical enterprise might produce 88,000 bricks a year and employ two workers for each firing. In La Huaca in Piura, the brickmakers use kilns with a capacity of between 7,000 and 14,000 bricks. An average capacity kiln, containing 12,000 bricks, measures about 2.80 m x 3.37 m x 4.00 m. A typical enterprise might produce 132,000 bricks a year and employ three workers per firing. Enterprises in La Compañía in the Huamanga province, department of Ayacucho, have kiln capacities of between 15,000 and 60,000 bricks. A typical enterprise would use kilns with a capacity 26,000 bricks, measuring around 5.00 m x 3.32 m x 4.50 m, and produce around 286,000 bricks a year. Typically, each firing employs four people. Finally, in Cerrillo, San José and Sultín in Cajamarca, kilns have a capacity of between 10,000 and 34,000 bricks. A typical capacity is 27,000 bricks and the kiln measures approximately 4.00 m x 4.50 m x 5.00 m. Such a kiln enterprise might employ 10 people per firing and produce 297,000 bricks a year.

Now, let us consider the nature of the produce of Practical Action's brickmaker partners. Solid 'King Kong' type bricks are the most common fired clay material used to build walls throughout Peru, and hence are the backbone of brickmakers' production. There is some variations from region to region and King Kong bricks in Tumbes measure 233mm x 116 mm x 81 mm, while in Piura the size is 208 mm x 120 mm x 83 mm, in Ayacucho 217 mm x 115 mm x 85 mm, and in Cajamarca 223 mm x 129 mm x 77 mm. The average mass of a solid fired brick is currently 3.59 kg in Tumbes, 2.76 kg in Piura, 3.15 kg in Ayacucho, and 3.14 kg in Cajamarca. One practical reason for noting this variation, particularly in mass, is to highlight the folly of comparing energy used per brick. Assuming the vitrification temperatures of soils and other factors to be equal, firing a certain number of bricks in Tumbes would be expected to consume 1.3 times as much energy as firing the same number to the same degree in Piura. Therefore we must compare specific firing energy, i.e. the energy used per unit mass of fired brick, e.g. kilojoules per kilogram (kJ/kg).

The National Technical Standard does not give absolute measurements for bricks, but rather the permissible percentage variations. Specified percentage variations in size apply to each and every dimension of the brick. Distortion is a measure of concavity or convexity, moreover. A Type III brick from Tumbes, for example, could vary in size, in whole numbers, from 226 to 240 mm x 112 to 120 mm x 77 to 85 mm. In terms of distortion, when such a brick is laid on a flat surface there should not be a gap of more then 6 mm between that surface and the end, middle or indeed any point of the brick surface. In terms of incorporating wastes into the body of the brick, it is important to note not only the compressive strength specification but also that for density.

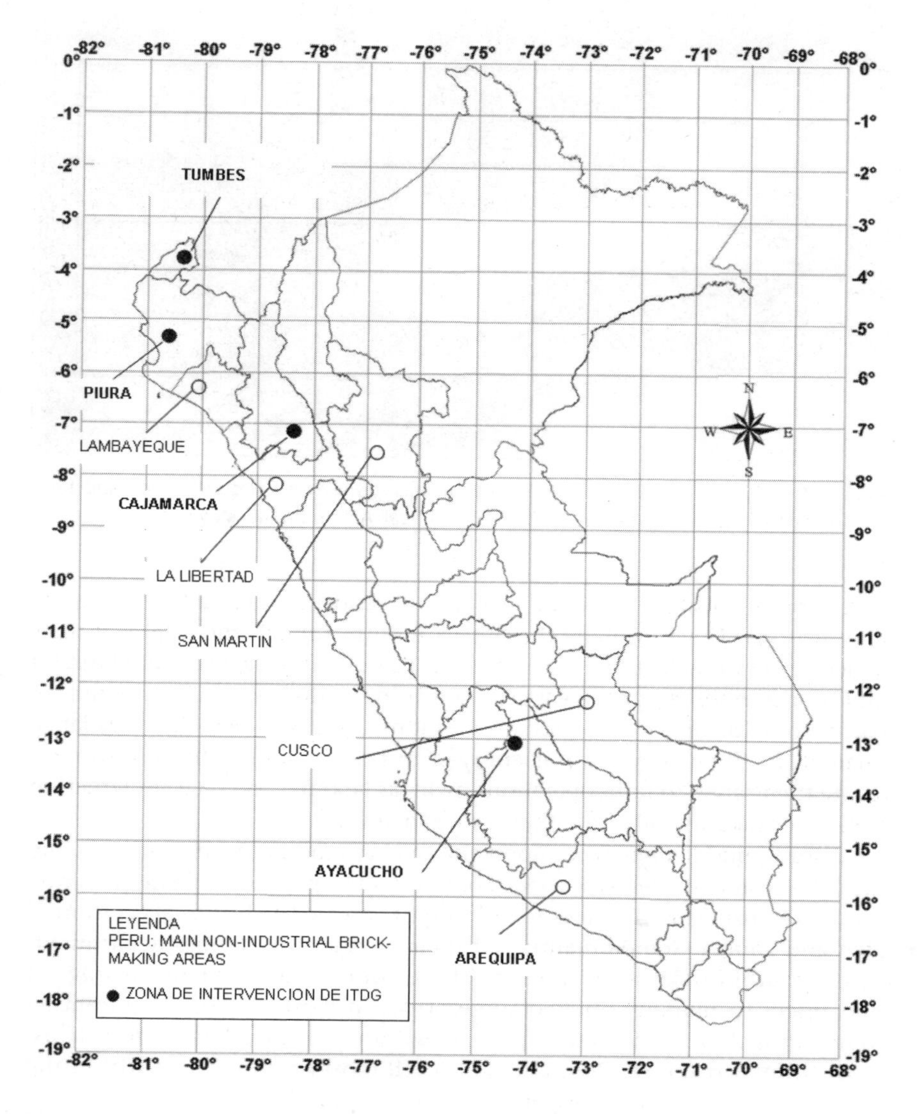

Figure 4.2 Brickmaking areas of Peru and locations of Practical Action's intervention
o Main non-industrial brickmaking areas
• Areas of Practical Action's involvement

Photo 4.2: A 12,000 brick Scotch Kiln at La Huaca owned by Enrique Estrada and fired with wood, sawdust and rice husks. Credit: Practical Action/Emilio Mayorga.

Brickmakers usually work all year round, reducing their activities during rainy seasons. They normally fire once a month, though the total is usually reduced to 11 times in a year due to rain. In Tumbes a typical kiln would run for only 8 or 9 months of the year. The operating period is similar in Piura. In Ayacucho, kilns operate for perhaps 10 months per year, which is similar to Cajamarca. When they are not engaged in brickmaking activities, workers devote themselves to farm operations, work as labourers on civil construction projects, or take any other job that comes their way.

Typically, middlemen, who supply the areas around production centres, buy bricks from small-scale brickworks. Sometimes the producers themselves hire trucks to transport their bricks to sell in other areas. The brickmakers from El Eden and Pampa Grande mainly sell their products in Tumbes department, typically within 10 km of their plants. Those from La Huaca in Piura mainly sell their bricks to markets situated some 30 km distant. Brickmakers from La Compañia in Ayacucho sell their produce within a 20 km radius of their plants. Those from Cerrillo, San Jose and Sultin sell most of their bricks within the Cajamarca province, supplying markets some 15 km distant.

Prior to Practical Action's involvement, at least, the bricks produced in La Huaca and in La Compañía were of only Type I, which according to the National Technical Standard are fit for masonry constructions under *minimum* service conditions, which means they have relatively low compressive strength and

Table 4.4 Calorific values of fuels and prices per GJ energy

Fuel	*Net calorific value (GJ/tonne)*	*Price per GJ (US$/GJ)*
Guayacán firewood	16.986	0.40
Carob tree firewood	16.794	1.10
Eucalyptus firewood	18.192	1.59
Sawdust	17.777	0.58
Rice husks	13.407	0.47 (Tumbes)
		0.00 (Piura)
Semi-bituminous coal	20.743	3.71
Anthracitic coal-dust	26.854	1.95
Waste motor oil	25.394	4.29
Diesel oil	40.399	23.45

probably, as is usually concomitant, low density and high water-absorption. In El Edén, Pampa Grande and Cerrillo no studies were conducted prior to Practical Action's involvement, but bricks made in similar rural areas at such scales of production would generally be either of Type I or, at best, Type II. Such bricks are not well suited to anything but the construction of single-storey dwelling walls and would not be much in demand in more lucrative urban markets.

In all areas of Practical Action's involvement, brickmakers use mainly Scotch Kilns. Following the technology-transfer initiative by Practical Action, brickmakers realize that clamps adapted from the coal-fired type tested by Practical Action in Zimbabwe are good for firing with rice husks. Due to a combination of circumstances, however, they only use them occasionally.

Innovations in fuel substitution

Prior to Practical Action's involvement, brickmakers in El Edén and Pampa Grande in Tumbes used Guayacán wood and some small proportion of rice husks as fuel (Mayorga, 1999). In Piura, carob tree wood and dung were used, while Ayacucho brickmakers used eucalyptus wood and semi-bituminous coal-dust. Eucalyptus is the fuel used by the brickmakers of Cerrillo, San José, Sultín-Álamo and Colcapampa in Cajamarca. Practical Action and its partners instigated trials with anthracitic coal-dust briquettes and waste oil burning in La Huaca, Cerrillo, San José, Sultín-Álamo and Colcapampa. In La Huaca, brickmakers also experimented with using sawdust as a fuel substitute. Comparing the cost of the various fuels and fuel substitutes, at 2006 prices (US$1 = S/3.25 (nuevo soles) in November 2005:

- In Tumbes, Guayacán firewood costs US$6.75/tonne, while rice husks cost US$6.32/tonne.

- In Piura, carob tree firewood costs US$18.46/tonne, while sawdust costs US$10.26/tonne and rice husks can be obtained free of charge at the local mill.
- In Ayacucho, semi-bituminous coal costs US$76.92/tonne, while eucalyptus firewood costs US$29/tonne.
- In Cajamarca, waste oil costs US$108.80/tonne, while diesel (for the motor to run the waste oil burner) costs US$947.30/tonne, and anthracitic coal-dust trucked in from La Libertad costs US$52.31/tonne.

In Tumbes there is relatively little financial incentive for brickmakers to consider agricultural waste as a substitute for a proportion of their primary fuel. In Piura on the other hand, both sawdust and rice husks offer brickmakers significant savings if they can technically be substituted for carob tree wood. The brickmakers of Ayacucho seem to be better off using eucalyptus than semi-bituminous coal, though the former is certainly not a cheap option when compared to wood and wastes in other regions. In Cajamarca, from a financial point of view anyway, brickmakers would be advised to fire with anthracite coal-dust. Waste motor oil does not offer any financial advantage as a principal fuel. Brickmakers obviously believe that its use to start a coal-fuelled clamp is beneficial, however, as they persist in this practice. If available, it seems Cajamarca brickmakers should consider using wood, sawdust and rice husks.

Typically, brickmakers pay nothing for their raw materials, clay, sand and even water. The proportionate cost of fuel varies depending on the location considered. Before Practical Action's involvement in Piura, for example, the only fuel used in the traditional technology was firewood, with dung used solely to cover the top of the kiln, i.e. principally as an insulator rather than a fuel. The fuel cost amounted to some 48 per cent of the total production cost. Current firing techniques have changed and brickmakers now use proportions of sawdust and rice husks in addition to firewood (APGEP-SENREM/ITDG, 2002). So, following Practical Action's involvement, fuel consumption had been reduced to between 38 and 43 per cent of total production costs. This saving has become critical to the survival of brickmakers in Piura because during the period we are considering the selling price of bricks reportedly fell by 20 per cent. It seems that this drop in price applies not just in Piura but virtually nationwide.

In Ayacucho, the cost of fuel is equivalent to 43 per cent of total costs. In Cajamarca, the traditional wood firing technology subsists side by side with the technology that Practical Action introduced in trials, wherein anthracitic coal-dust briquettes are used as the main fuel with waste burned to light the kiln. In the wood firing technique, the cost of fuel is equivalent to 17 per cent of the total cost. With the use of coal-dust and waste oil, fuel consumption rises to between 21 and 24 per cent of total costs. Surely, though, brickmakers must perceive a benefit in terms of product quality, productivity or fuel availability or they would not persist with the innovation. Though there may be such benefits,

Box 4.2 A brickmaker's tale

Juan Francisco Coronado Acaro was born in Viviate in La Huaca district in 1933. Like the majority of people in Viviate, his first job was helping to make brooms and straw mats. In due course, he began producing and selling brooms for himself, travelling to markets far and wide to sell his wares. When the advent of plastic brooms rendered his craft obsolete, Juan had been involved in this business for nearly 25 years. In 1987, a brickmaker friend invited him to become a partner and so he began brickmaking at the La Huaca site in Piura. Eventually, Victor Carmen, a leading brickmaker in the area, sold him a 14,000 capacity brick kiln, which Juan paid for in instalments as the money from sales came in. Between 1990 and 1992, Juan worked as a Councilman in the sports and culture division of the District Council of La Huaca. He is well remembered for his efforts in organizing the football league.

From 1993, Juan devoted himself to his brickmaking business. He built another 7,000 brick capacity kiln that he used for trial purposes. During the time that Practical Action worked in La Huaca, Juan was elected to the Board of Directors of the Brickmakers' Association. An enthusiastic project partner, he went on an Practical Action sponsored visit to Trujillo to visit coal suppliers, workshops where waste oil burners are made, and brickmaking enterprises. Entrepreneurial and a natural innovator, Juan was the first brickmaker in La Huaca to experiment with the use of sawdust in the clay mix, enlarging his moulds to solve the problem of increased drying shrinkage. Moreover, Juan was the main user and promoter of oil burner technology in the area. He died of a heart attack in September 2002 age 69, while negotiating the sale of bricks.

in technology transfer projects psychological factors can no more be ignored than they can in other aspects of life. Oil-burning technology is much used in Trujillo, La Libertad, where specialized entrepreneurs provide the service to brickmakers. Trujillo is known as a source of good quality bricks and an area where larger brickmaking enterprises thrive. So, the benefit of oil-burning technology in Cajamarca may be that it is perceived as modern or hi-tech and therefore boosts brickmakers' prestige and/or that of their products.

Let us take a closer look at the fuel supply situation and any problems brickmakers face in attaining their fuel of choice. Although it is forbidden to fell trees for firewood, the government's capacity to control such activities is virtually non-existent. According to most brickmakers, they have no problem obtaining firewood. One exception to this rule seems to be around La Huaca, where conservation legislation is somewhat more rigorously enforced. Rice husks are waste material and are available free of charge at many local mills. Nevertheless, the cost of transporting them restricts their use. In La Huaca, there is a mill nearby and brickmakers collect the husks themselves. While this is cheaper than having the husks delivered and, importantly for cash-strapped brickmakers, does not involve a direct payment in advance of making bricks, there is obviously still a transport cost comprised of labour, fuel and vehicle overhead costs. Brickmakers do not tend to consider this 'hidden' cost in assessing their production costs, however. The long-term, global-scale, contributory environmental impact of transport is certainly not a significant consideration

Photo 4.3 A 34,000 brick kiln in Cajamarca, owned by the Old People's Home, which uses coal-dust briquettes (shown) as a fuel. Credit: Practical Action/Emilio Mayorga.

for brickmakers in La Huaca: their margins are tight, livelihoods perilous and thus income is their overriding concern.

Brickmakers in Tumbes, by contrast, have to pay the cost of hiring a truck and so that cost is formalized, apparent and considered. Coal-dust is a residue that is always on sale in coal production areas. Although it can be transported to the desired areas without any problem, the cost of transport increases the price significantly in proportion to distance. It is a similar story for sawdust, which is in large measure a waste material and usually available at minimum cost if there are sawmills in an area. Waste motor oil, which has been used for reprocessing as well as for fuel in recent years, is nevertheless still in constant supply nationwide. Diesel is available from local service stations.

Methods of using the various fuels and their distribution in the kiln vary. Sawdust, for example, can be burned in fires at the base of the kiln or, more usefully, incorporated in appropriate proportions into the body of clay bricks at the forming stage. Post Practical Action's intervention many brickmakers in Piura have substituted sawdust and/or rice husk for a proportion of firewood and the use of dung has been virtually abandoned. Firewood is used at the base of the kiln, sawdust is used in the clay mix and rice husks are used to cover the kiln. Problems persist with the quality of bricks, however, as brickmakers experiment with, particularly, finding the maximum amount of sawdust that it

Energy consumption of brick firing process

NAME OF PRODUCER	LOCATION/ADDRESS	DATES OF FIRING
Victor Carmen	La Hauca, Paita, Piura	Start: 17 May 1997 at 15:45 Finish: 18 May 1997 at 17:00
TYPE OF CLAMP/KILN	TYPE(S) OF FUEL	MASS OF FUEL(S) USED (kg)
2 tunnel, Scotch Kiln 3.25 m x 2.45 m x 3.3 m	Algarrobo wood Semi-bituminous coal-dust	Algarrobo: 2,270 Coal-dust: 1,400
CALORIFIC VALUE(S) (kJ/kg)	NO. OF GREEN BRICKS	AVG. MASS OF BRICKS (kg)
Algarrobo (net): 16,310 Coal-dust (net): 15,547	6,358	Wet = 4.20 Dry = 4.11 Fired = 3.75
BRICK MOISTURE CONTENT	METHOD OF FORMING	WEATHER CONDITIONS
2.14%	Slop moulding	Hot & dry with light gusting wind
CALCULATION OF KILN EFFICIENCY		QUALIFYING INFORMATION
Mass of green brick = 26,704 kg Total moisture content = 572 kg Drying energy = 1,482,622 kJ Wood energy = 37,023,700 kJ Coal energy = 21,765,800 kJ Gross energy = 58,789,500 kJ Firing energy = 57,306,878 kJ Mass of fired brick = 23 843 kg Specific firing energy = 2.40 MJ/kg		(i) Vitrification temp of soil = 1,150°C (ii) Max kiln temp = 970°C (iii) Firing time = 38h 15m

COMMENTS

Small kin and therefore likely to be inefficient. Initial firing too rapid to take advantage of placing coal in layers. Bricks produced: approximately 90% good, 10% under-fired and broken.

NAME, CONTACT DETAILS & DATE
Emilio Mayorga, ITDG Peru, 31 October 1997

Figure 4.3 Energy monitoring form

is technically feasible to incorporate. The relatively low cost of sawdust together with the problem of fuelwood supply tempts brickmakers to overuse the former and risk producing bricks that do not even meet the requirements of the market for the lowest quality, Type I, bricks. Meanwhile, the innovation of oil-burning technology at the La Huaca factory has been abandoned due mainly to the untimely death of Juan Francisco Coronado Acaro.

In Cajamarca, Practical Action's project has resulted in brickmakers adopting the use of anthracitic coal-dust and waste oil. Briquettes are prepared with coal-dust, clay and water. Once dry, these are placed on grills made with parallel rows of bricks at the base of the kiln. Bricks are loaded into the kiln to form arches over these grills. Subsequent layers of bricks are loaded, with coal-dust evenly distributed between each layer, until the kiln is full. In addition, a number of briquettes are placed among the bricks along the edges of the kiln, where bricks tend to be under-burned in a conventional firing. An oil burner, burning waste engine oil and driven by a diesel fuel motor, is used to light the briquettes at the base of the kiln. After this, the burning process proceeds unassisted, apart from perhaps regulating the air intake and mitigating wind effects.

In Tumbes, the technique of burning rice husks developed by Practical Action and its partners is used during the rainy season, when the poor state of the roads restricts the supply of firewood. Rice husks are placed between parallel walls formed by green (unfired) bricks and also to cover the top of the Scotch Kiln or clamp, where they act both as fuel for the top layers of bricks and as thermal insulation. The technique of placing rice husks between parallel walls of green bricks was developed by brickmakers in La Huaca. The process of technology transfer engaged in by the Practical Action project team resulted in it being replicated by brickmakers in El Edén and Pampa Grande in Tumbes to solve their seasonal problem with fuel supply. In dry seasons, brickmakers persist with their familiar technology, whereby firewood is used at the base of the kiln and rice husks are incorporated as fuel in the clay mix as well as to cover the kiln. Since Practical Action's involvement, however, brickmakers have experimented and are now able to incorporate larger volumes of rice husk in the body of bricks without undermining product quality or sale value.

In Ayacucho the technology that was in place before the Practical Action project persists. Firewood is used at the base of the kiln and semi-bituminous coal-dust between layers of bricks. It is often the case that local technologies, which have evolved over time and through experience, are the optimum solution and the Practical Action team were acutely aware of this when instigating projects. In some instances, though, a fresh look *will* reveal opportunities for innovation.

Let us consider how much fuel is used to fire typical kilns. In an 8,000 brick kiln in Tumbes, using guayacán firewood in the base and rice husks in both the clay mix and to cover the kiln, the mass of fuel used is 10,809 kg of firewood plus 828 kg of rice husks. A 12,000 brick kiln in Piura uses 3,400 kg of carob tree firewood at the base, 3,600 kg of sawdust in the clay mix and 391 kg of rice husks as the cover. In a 26,000 brick kiln in Ayacucho, 8,000 kg of eucalyptus firewood is burned at the base and 2,000 kg of semi-bituminous coal-dust between the layers of bricks. A 27,000 brick kiln in Cajamarca, employing a waste fuel oil burner to light coal-dust briquettes at the base and coal-dust between the layers of bricks, uses 3,500 kg of anthracitic coal-dust, 198 kg of waste oil, and 3.2 kg of diesel to fuel the oil burner motor.

From the data we now have for fuel consumption and calorific values, we can calculate how much energy is used in firing processes at the various locations. Though this is of some interest, it does not tell us enough to compare the processes and say which is most efficient. What we really want to know is not only the total energy used, but also the amount of that energy used to complete the drying of bricks in the kiln and the amount actually used in firing – vitrifying

Table 4.5 Summary of energy data for the locations considered

	Tumbes	*Piura*	*Ayacucho*	*Cajamarca*
Total energy (MJ)	194,701	126,337	189,268	99,151
No. of bricks in kiln	8,000	12,000	26,000	27,000
Brick moisture content (kg)	1,378	1,582	2,600	3,774
Drying energy (MJ)	3,571	4,099	6,737	9,868
Drying energy/total energy (%)	1.83	3.25	3.56	9.96
Firing energy (MJ)	191,131	122,238	182,531	89,283
Specific firing energy (MJ/kg)	7.32	3.63	2.19	1.15

– the bricks. From that data we can then calculate the *specific firing energy* in kilojoules per kilogram (kJ/kg) of fired brick. This figure for specific firing energy allows the performance of any brick firing process to be directly compared with any other. The methodology for calculating and recording specific firing energy is detailed in Chapter 7 of *Brick by Brick* (Mason, 2001). In Figure 4.3 a summarized example of a completed data collection form is reproduced.

In Table 4.6 the data for the four locations and typical conditions we have been considering are summarized. It is still not quite the full story, however. The table tells us nothing about the environmental impact of the processes compared. Neither does it, either quantitatively or relatively, detail the quality of the bricks produced, though there is room in the comments section for a qualitative assessment. To illustrate the strengths and weaknesses of the methodology, consider Cajamarca. One reason the specific firing energy is low in that instance may be that bricks are significantly under-fired. Another reason, which is evident to the experienced eye, is that the typical kiln in Cajamarca has the largest capacity and so is almost certainly the biggest of those considered. All else being reasonably equal, therefore, the Cajamarca kiln is likely to be most efficient because bigger cubic kilns generally have a lower surface area to volume ratio and so lose less heat (Mason, 2000b).

We probably have enough data now to calculate which location is employing the most cost effective means of firing. There is no practical point, however, as the technologies employed in the four locations have been developed as the best possible in each context, i.e. the choice of fuels and kiln size in Cajamarca are not realistic, everyday options for brickmakers in Tumbes. From Table 4.6, however, if brickmakers in Tumbes did want to save energy, it is apparent that they should consider using bigger kilns. Meanwhile, brickmakers in Cajamarca could cut their fuel costs by drying bricks more thoroughly before they are placed in the kiln. As a general guide, a specific firing energy of 2 to 2.50 MJ/kg of fired brick is reasonable to expect with good practice in small-scale production. Lower values for specific firing energy may indicate that bricks are under-fired.

Photo 4.4 José Morán's 6,000 brick clamp in Pampa Grande, fuelled solely by rice husks. Credit: Practical Action/Emilio Mayorga.

Practical Action's interventions revisited

Chapter 9 of *Brick by Brick* summarizes project outcomes at the end of ITDG's active involvement with brickmaking projects in Peru in 2001. In general, brickmakers have significantly improved their capacity to adopt and adapt technologies in accordance with market demand. The energy monitoring methodology and participative ways of working were also successful project outcomes. The environmental imperative, 'more to be done', was signposted. However, three technologies were earmarked as worth further consideration, perhaps development and wider dissemination:

- incorporating sawdust and rice husks as fuels in bricks;
- hand moulded (low pressure) briquettes of coal-dust and clay;
- waste oil burner (with the environmental cautions noted).

Practical Action Peru revisited project partners in July and August 2005. The purpose was to make a rapid appraisal of how technological innovations had fared and what the longer-term impacts of projects had been on livelihoods. In El Edén and Pampa Grande, trained brickmakers have gone back to using firewood and their Scotch kilns. As a result of a visit they had made as part of the training provided by Practical Action, however, these brickmakers have modified their kilns to resemble those used in La Huaca. The modification consists of eliminating the fixed adobe arches built on the bottom of the kiln, in which they would place the firewood. Instead, they now build false channels with green bricks that are to be fired. This increases the effectiveness of the kiln and saves fuel. Brickmakers also use a larger proportion of rice husks in the clay mix than they did before Practical Action's involvement.

In Pampa Grande, one of the trained brickmakers has built a Scotch Kiln fired only by rice husks. This type of firing is only used during the rainy season when, due to shortage of supply, the market does not insist so strongly that bricks are the deep red colour normally required. It is worth mentioning that when Practical Action provided training at the end of its involvement (October–December 2000) it happened to be the rainy season, therefore using rice husks as the sole fuel was initially accepted very enthusiastically. Once the three- or four-month rainy season was over, however, brickmakers resumed their traditional technology in order to satisfy the colour preference of their customers.

In La Huaca, brickmakers are currently using Scotch Kilns, employing carob tree firewood for external firing, sawdust and rice husk ash mixed in the brick clay, and rice husks to cover the kiln. The bricks produced in La Huaca are of a poor quality with a very low bearing capacity that does not meet even the lowest requirement of the National Technical Standard. Comparing current practice with firing using only wood at the base of the kiln reveals a 28 per cent saving in fuel costs. This is despite the fact that current practice consumes somewhat more energy, between 3.63 and 4.16 MJ/kg, than did the traditional technology at a measured 3.40 MJ/kg. Moreover, it consumes much more energy than did using coal and a waste oil burner in trials between 2000 and 2002, i.e. only 1.58 MJ/kg. As mentioned earlier, firing with waste oil was abandoned at La Huaca about three years ago, upon the death of Juan Coronado. Apart from the loss of Juan's energy and enthusiasm, a measure of superstition has mitigated further trials with the oil-burning technology that he was so strongly associated with; it has become taboo.

In Ayacucho, Type I bricks are produced. The brickmaker in La Compañía trained by Practical Action, resumed using 'traditional' technology two years ago, burning firewood and coal-dust in his Scotch Kilns. This was due to a shortage in the supply of waste oil, which is now used as a fuel for cooking food on a large scale in Ayacucho. It is also used for reprocessing purposes in Cajamarca and to produce substandard fish-meal in clandestine factories in Piura and Tumbes. Fish-meal or fish-feed, manufactured from the Peruvian anchoveta, is exported to feed farmed salmon for the tables of Europe and the USA. As a consequence of demand, anchoveta is over fished in Peruvian waters. Moreover, PCB contamination in fish-meal and fish-feed, and thence salmon for human consumption and so humans, is a growing concern.

Compared to other areas, there has been a longer lasting assimilation of the results of Practical Action's intervention in Cajamarca. In Cerrillo, San José, Sultín-El Álamo, and Colcapampa, non-industrial brickmaking enterprises continue to employ the technology piloted by the Practical Action project. They use a waste oil burner, coal-dust briquettes and coal-dust to fire the bricks. The cost of this practice is similar to using firewood but saves a significant amount of time and labour. Firing 10,000 bricks now only takes three hours of active firing, for example, instead of the two and a half days required formerly, because, once alight, the process is largely self-promoting. Environmentally, the main positive impact is the 100 per cent reduction in the use of carob tree wood. The environmental impact of this new technique has not, however, been compared with traditional practice. While there will have been a positive impact on deforestation, emissions of harmful substances that cause a variety of negative environmental effects would appear to have been increased. The majority of bricks produced in Cajamarca are of a poor quality that does not meet the requirements of the National Technical Standard, although Type I bricks are also produced.

Following Practical Action's involvement, brickmakers in both Tumbes and Piura employ a larger volume of rice husks than in their traditional firing techniques. They are also aware of the firing technique that employs only rice husks. Burning additional rice husk does seem to offer a net environmental benefit. Firstly, less wood or coal is used, thereby reducing deforestation or emissions associated with coal burning. Moreover, as the means of disposal of rice husks, deemed a waste product, would anyway be burning or rotting, overall emissions of carbon monoxide and carbon dioxide are most probably reduced. Another technological innovation was noted. The Practical Action project team had investigated adding ashes to the clay mix to reduce the accumulation of ashes after firing bricks with rice husks as fuel, a landscape and atmospheric pollution that makes brickworks unpleasant and less healthy places to work. This technological innovation to dispose of waste and clean up the site was successful and is currently being used by some of the brickmakers trained by Practical Action in La Huaca, El Edén and Pampa Grande.

In conclusion, it is worth stressing some realities of the situation with respect to energy use and the environment. Generally, brickmakers will readily use protected natural resources as fuel if they can get away with it and increase their income. Moreover, they choose the firing technologies that ensure them the greatest income as long as can produce the bricks that the market demands. If the market will accept poor-quality bricks that have a limited bearing capacity, then that's what brickmakers will produce. Buyers in the market segment satisfied by non-industrial brickmakers in Peru are not usually discriminating about the quality and bearing capacity of the bricks they purchase. Essentially, they consider anything that holds together to be acceptable for construction, and they judge this mainly by the weight of the brick. Otherwise, their predominant selection criteria are colour and, of course, cost. For non-industrial brickmaker in Peru, then, securing a day-to-day livelihood is such a precarious business that the wider and future environment is a distinctly low priority.

CHAPTER 5
The use of cow-dung, bagasse and a variety of other agricultural residues in Sudan

Ahmed Hood

Land, climate and people

Populated by an estimated 34.5 million people, Sudan is located in the north-eastern part of Africa. It has land borders with Egypt, Libya, Chad, the Central African Republic, the Democratic Republic of the Congo, Uganda, Kenya, Ethiopia and Eritrea. It also faces Saudi Arabia across the Red Sea. Covering some 2,505,810 km^2, more than ten times the area of the UK, Sudan is the largest country in Africa.

The physical geography of Sudan is dominated by the River Nile and its tributaries, the White and Blue Niles. In the central region, the landscape is generally flat, featureless plain. By contrast, mountains dominate in the north-east, west and far south, where they rise to over 3,000 metres. The northern region is mainly desert. Nationwide, the climate ranges between tropical in the south and arid desert in the north. The overall climatic picture encompasses a variety of microclimates, ranging from humid tropical, through temperate Mediterranean, to desert. The rainy season varies by region and rainfall varies between less than 200 mm in the far north to 1,500 mm in the far south.

Table 5.1 Selected demographic and social indicators

Population growth rate (1998/2003) (%)	2.63
Urban population (% of total pop., 2003)	35.52
Population under 15 years of age (% of total pop., 2003)	42.04
Population 60+ years of age (% of total pop., 2003)	3.94
Crude death rate (per 1,000 pop.), 1998/2003	11.50
Infant mortality rate (per 1,000 live births), 1999	68.00
Average household size (persons), 1999	6.40
Literacy rate (age 15–24)	54.80

Source: CSS, 2004.

Figure 5.1 Map of Sudan

Economics in brief

To a large extent, Sudan has turned around a struggling economy via a set of economic policies and infrastructure investments. The nation still faces formidable economic problems, however. A key cause is reportedly the low level of per capita output. From 1997 to date, Sudan implemented a programme of International Monetary Fund (IMF) macro-economic reforms. In 1999, it began exporting crude oil, and in the last quarter of that year recorded its first trade surplus. This factor, along with monetary policy, has stabilized the exchange rate though the rate of inflation remains volatile. Gross Domestic Product (GDP) is US$14,956 million. Increased oil production, revived light industry and expanded export processing zones helped sustain GDP growth at 6.4 per cent in 2004. Agriculture remains a critical sector of the economy, employing 80 per cent of the workforce, contributing 39 per cent of GDP, and accounting for most GDP growth. The majority of farms depend directly on seasonal rainfall, however,

Table 5.2 Production of major crops in Sudan ('000 tonnes)

Year	*Cotton*	*Peanuts*	*Sesame*	*Sunflower*	*Sorghum*	*Millet*	*Wheat*	*Gum Arabic*
1998/99	165	776	262	10	4,284	667	172	-
1999/00	147	1,047	329	8	2,347	499	214	10.6
2000/01	232	947	282	4	2,491	481	303	24.4
2001/02	243	990	269	4	4,394	578	247	25.1
2002/03	254	555	122	19	2,875	581	331	22.9

Source: CBS, 2004.

and are consequently highly susceptible to drought. Chronic instability ensures that much of the population of Sudan will remain at or below poverty line for many years to come. Causes of this instability include the long-standing civil war, adverse weather patterns and weak world agriculture prices.

Until the new millennium, agriculture was the economic backbone of Sudan. Livestock farmers tend around 135 million head of sheep, goats, cattle and also camels. Major crops are sorghum, millet, wheat, cotton, groundnut, sesame and sunflower. While sorghum, millet and wheat are important staple foodstuffs, the other crops are grown mainly for export. Another important export is gum arabic. With such a large percentage of the population dependent on livestock and arable farming for their livelihoods, prosperity, sufficiency and even survival hinge precariously on the rains. For at least the last three decades Sudan has been hit by repeated and unpredictable droughts that have had a disastrous effect on livelihoods throughout the nation. There is widespread poverty and levels of internal and external migration are high. A complex relationship exists between drought, desertification, poverty, an itinerant population, the ongoing civil war, and tribal conflicts in some regions.

Before the advent of petrol production, Sudan had a meagre industrial sector based mainly on processing agricultural produce. Dominating this sector were sugar cane processing and the textile industry. In recent years, however, the textile industry has undergone economic and technical problems and has all but closed down. Other industries include the production of cement, vegetable oil, soap, leather and flour. In tune with IMF prescriptions, the industrial sector is oriented towards exports rather than import substitution or sufficiency; dependency and not frugality are the order of the day. Similarly, the growing mining sector is geared towards the export of unprocessed ores and minerals, rather than value-added products. In 2003, the nation produced 5,106 kg of gold, 2,844 kg of silver, 15,000 tonnes of chrome and 13,304 tonnes of gypsum.

Energy

Since 2000 Sudan has been an oil producer and exporter. The total current annual production was around 350,000 barrels per day and was expected to reach 500,000 barrels by the end of 2005. From the figures and tables included in this

Table 5.3 Primary energy consumption ('000 toe)

Energy source	*1980* *Quantity*	*%*	*1999* *Quantity*	*%*	*2003* *Quantity*	*%*
Petroleum products	1,147.1	16.1	1,666.0	15.41	2,300.0	20.3
Electricity (hydro)	63.3	0.9	104.2	0.96	226.0	2.0
Biomass	5,909.6	83.0	9,040.4	83.63	8,800.0	77.7
Total	7,120.0	100.0	10,810.6	100.00	11,326.0	100.0

Source: Sudan Energy Assessment. Contact Practical Action Sudan for further data.

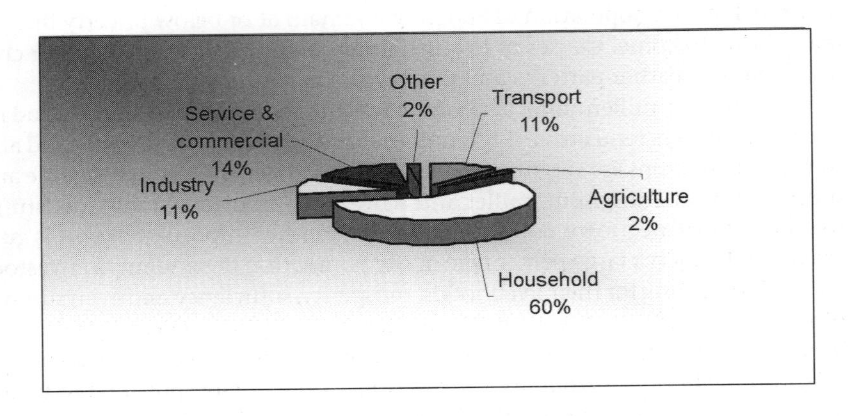

Figure 5.2 Energy consumption by sector
Source: Contact Practical Action Sudan for further data.

section, however, it is evident that domestically Sudan depends heavily on biomass energy sources: firewood, charcoal and agricultural residues. Of the 78% primary energy consumption accounted for by biomass in 1999, some 69% was woody biomass and 9% residues. There is a geographic mismatch between areas of biomass production and distant centres of its consumption. With biomass transported for distances of over 1,000 km, there is a significant burden on the transport system, with the concomitant environmental impact. Obviously, the financial, if not environmental, cost of transport is reflected in the selling price of biomass. Moreover, with Sudan facing environmental degradation due to deforestation, drought, over-grazing and subsequent desertification, dependency on woody biomass is problematic, particularly as the vast bulk is attained from forestry that is not sustainably managed.

The household sector accounts for about 60% of total energy consumption and this is mainly biomass. In fact, the sector consumes some 72% of total biomass energy and about 50% of electricity. Biomass is burned in traditional inefficient stoves, with extensive smoke emission that constitutes a health hazard

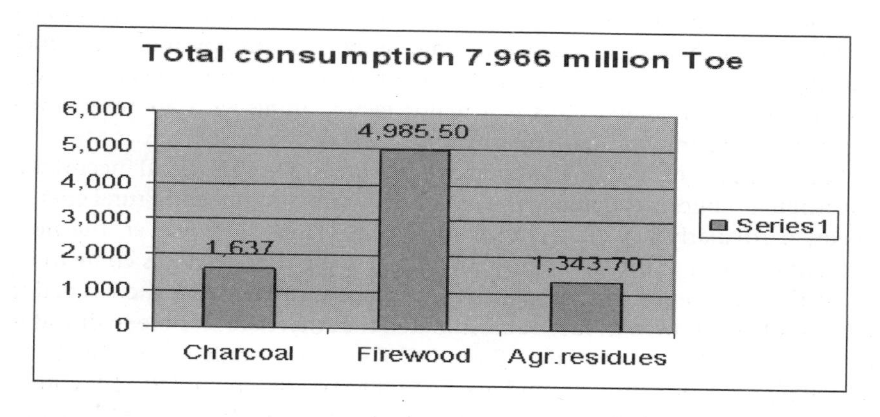

Figure 5.3 Biomass energy consumption, 1999

in poorly ventilated dwellings. Attempts to introduce improved stoves have not been very successful. Similarly, the use of agricultural residues as domestic fuels has not caught on. A combination of the initial financial outlay and resistance to change are believed to be the reason for the continuing inefficient use of biomass. The high rate of urbanization means increasing consumption of biomass in cities, moreover. Charcoal, which is lighter to transport and cleaner to burn in the home, is the subject of particularly strong demand. Though the use of liquified petroleum gas (LPG) in urban areas has increased, once again the high cost of appliances and adherence to traditional cooking methods limit its uptake.

Table 5.4 Consumption of biomass by sector, 1999 ('000 toe)

Sector	*Charcoal*	*Firewood*	*Agricultural residues*	*Total*	*%age of total*
Household	1,456.6	3,553.5	732.0	5,742.1	72.1
Services	179.0	1,069.7	-	1,248.7	15.7
Industrial	1.2	362.3	611.7	975.2	12.2
Total	1,636.8	4,985.5	1,343.7	7,966.0	100

Source: Practical Action/ITDG Sudan. Contact Practical Action Sudan for further data.

Table 5.5 Biomass energy consumption by industry scale, 2001 (tonnes)

Biomass fuel	*Small-scale*	*Medium-scale*	*Large-scale*	*Total*
Firewood	815,189.70	27,615.00	N/A	842,804.70
Agr. residues	126,828.70	6,706.50	2,313,359.60	2,446,894.80

Source: National Energy Assessment, Ministry of Energy and Mines. Contact Practical Action Sudan for further data.

Industry is suffering from similar stasis with respect to technology. Traditional industries, notably brickmakers and lime producers, employ inefficient fuel burning methods. And, once again, the financial investment needed to replace, for instance, dilapidated kilns is a major factor, along with a conservative approach to technology, particularly in rural areas.

Sudan has largely untapped renewable resources. Through fiscal incentives, government policies encourage the use of renewables for meeting rural energy needs, particularly via solar photovoltaic cells. There is, however, the now familiar problem of initial investment costs. For a nation that has come to be associated so strongly with drought, it is perhaps surprising that there is major scope for hydropower. In fact, it is estimated that there are feasible hydro sites with the potential to generate 4,860 MW. The total installed capacity of the national grid is presently around about 1,000 MW, of which 340 MW comes from hydro and 660 MW from thermal power stations. A 1,250 MW hydro scheme is under construction. Meanwhile, agro-industrial residues, including bagasse, are not exploited for electricity production, and the potential of cogeneration is largely unexplored.

Demand for electricity is highly suppressed, with only around 20 per cent of the population connected to the grid. All industries have to depend on standby generation. The short-term plan is to increase grid capacity by 580 MW from conventional thermal generation. In the medium term, significant investment will be required to further increase capacity by a planned 4,500 MW.

Traditional brickmaking

Brick production is a dry weather activity that is typically suspended for the approximately three-month rainy season. The industry is dominated by traditional technology, wherein the soft-mud or slop-moulding process is practised. There are only three modern brick factories and these do employ the stiff-mud or sand-moulding processes. At present only one of these modern factories is operational, however. High production costs mean that modern factories cannot presently compete with traditional labour-intensive brickmaking. Apart from the implication that labour and fuels are cheap and available to traditional brickmakers, there does not seem to be sufficient market for bricks of better quality.

In the soft-mud process, clay is dug manually and organic matter, particularly cow- dung, is widely used as an additive in almost all parts of the country. Due to the expansion in brick production and a scarcity of cow-dung in recent years, other organic wastes are now used as additives. In Kassala, eastern Sudan, bagasse from the New Halfa sugar factory is commonly used. In western Sudan, peanut shells are added to brick clay, particularly in enterprises sited around the city of Al Ubayyid. Other organic wastes used in different parts of the country include: wheat straw, sawdust and by-products of millet and sorghum threshing. Whenever cow-dung is available it is preferred over other organic wastes, however. This is because its inherent properties mean cow-dung improves the

Photo 5.1 Millet threshing waste as used in brickmaking. Contact Practical Action Sudan for further data.

workability of the brickmaking clay, reduces its fermentation (tempering) time, and cuts firewood consumption. Other organic wastes demand long fermentation time, as well as screening and grinding in some cases, before they are sufficiently broken down to allow moulding of the clay.

The amount of water used varies from place to place, depending on the type of clay and the particular inherited practices of the workers there. In general, it tends to vary between 30 and 40% of the mass of the clay. According to Hamid, for instance, it is about 33% of clay mass in Al Ubayyid and about 30% in the Khartoum area (Hamid, 1994). Once water has been added to the clay and additive mix, it is thoroughly blended with hoes and spades. The next stage of mixing is treading with bare feet, whence stones can be detected and removed. The mixture is then left to temper for about twelve hours. Then, before moulding, a final mixing operation is undertaken.

The most commonly used moulds are two-brick compartment steel moulds, open at top and bottom. These moulds are set on a removable wooden pallet that forms the base during moulding. In theory, for a typical clay, the dimensions of each mould compartment should be 220 mm x 110 mm x 60 mm. Such a mould is designed to yield a fired brick 200 mm x 100 mm x 50 mm. In practice, though, the dimensions of moulds differ from one production unit to another. Even within the same production unit, it is quite common to find moulds with different dimensions being used. Obviously, this leads to the production of

Photo 5.2 Cow-dung drying before being added to brick clay. Contact Practical Action Sudan for further data.

bricks of different sizes and reflects a thoroughgoing absence of standardization of practice in the industry. Variation in brick dimensions means bricklayers spend more time and use more mortar when building with the bricks. If standards were applied in brick production, moulds would be adjusted for each type of clay, to allow for drying and firing shrinkages, so that bricks of acceptable dimensional variance resulted nationally.

Nominally, the brick moulding process is carried out by five workers. Two deliver the clay mix from the tempering pit to the moulding site. The moulder extracts a quantity of clay that slightly exceeds the volume of mould. The clay is then thrown into the mould and the surplus removed by hand. Two types of moulding practice are commonly employed in Sudan. In one, a brick table is built and the moulder stands at and works upon this. The alternative is that the moulder stands in a hole and moulds bricks on the ground. In either case, once bricks are moulded, two workers carry the full moulds to a cleaned and flattened drying area. Here they de-mould the bricks, clean the mould and pallet in water if necessary, and return them to the moulder. The size of brickmaking enterprises is often measured by the number of such teams operating.

Bricks are dried in the sun. Freshly moulded bricks are laid out individually on their largest face. After 24 hours the bricks are dry enough to be turned over on to an edge. They are left for another one or two days to ensure uniform drying. The 'green' bricks are then stacked with spaces between them to allow air circulation and ensure leather-hard drying. Whether or not bricks are actually this dry before firing very much depends on the overall production process. It is common practice for the batch of bricks moulded first to undergo a drying period of 18 to 25 days, while the last batch made receives as little as 2 days' drying. Using the firing process to dry green bricks is, of course, an inefficient use of fuel and results in extra pollution and greenhouse gas emissions.

The only firing technology widely used by traditional brickmakers in Sudan is the brick clamp. Bricks are stacked in a truncated pyramid with firing tunnels built in at the base. Once built, the clamp is plastered with mud to seal it and act as thermal insulation. At the top an area is left unplastered to provide draught for the fire. Throughout the country, the principal fuel burned is wood. Over the last two decades, the supply and price of fuelwood have become major production constraints for brickmakers. So, these days, complementary fuels are often used in order to cut down on fuelwood use. Such complementary fuels are usually waste organic materials: animal dung and agricultural residues. Whereas these were once incorporated principally for their effect on clay properties, particularly reducing shrinkage and drying cracks in high-clay brick soils, wastes are increasingly being valued as fuels. Though this has positive benefits for livelihoods and the environment, the overuse of organic additives has contributed to a substantial decrease in brick quality, critically strength and water absorption.

Photo 5.3 A five-tunnel brick clamp. Contact Practical Action Sudan for further data.

Clamp firing involves charging the firing tunnels with wood. Once ignited at one end, the fire burns through the tunnel with little control. The fire then moves upwards through the clamp, igniting the complementary fuel that is either incorporated into the body of bricks or distributed throughout the clamp. Until the fire reaches the top of the clamp, experienced brickmakers control the amount of wood that is fed in to recharge the fires in the tunnels. When the fire has reached the top, both the tunnel openings and the chimney area are sealed, and repairs are made to any cracks in clamp plastering. The maximum temperature attained in Sudanese clamps is about 950°C, but not all clamps reach this temperature. Moreover, even in a clamp where this maximum is reached most bricks would experience significantly lower temperatures (Hamid, 1994). Depending on the nature of the clay and the experience of brickmakers, active clamp firing takes around 24 hours. The clamp is then left to 'soak' for about two days. After this time, the tunnels are opened up to permit more rapid cooling. When the clamp is finally dissembled, after perhaps a week or more, fired bricks are typically sorted into first-class, second-class, over-fired and under-fired grades.

In 1996, the Biomass Energy Network of Sudan, BENS, conducted a study of brickmaking at El Gereif. Situated in the Blue Nile valley, Khartoum state, El Gereif is a major brick production area. By no means statistically valid, the study reveals at least some degree of consistent quality at this particular site.

Previewing some of the discussion that will follow in the relevant section, Practical Action Sudan conducted a small study on the influence of cow-dung and the type of moulding process on the quality of brick made from Kassala clay (Bairiak, 1998). From Table 5.7, for slop-moulded bricks water absorption generally increases with the percentage of dung used. As might be expected, density inversely reflects the increase in water absorption, falling as it does in some proportion to the percentage increase in dung used. Results for compressive strength are not so clear. An initial dip in strength seems to reverse once more than 30 per cent of dung is added. The pattern is similar for sand-moulded bricks, with perhaps an even more confusing non-trend in results for compressive strength.

Table 5.6 Properties of bricks fired at El Gereif, Khartoum

Clamp No.	*Additive*	*Dimensions (mm)*	*Water absorption (%)*	*Compressive strength (kg/cm²)*
1	Cow-dung	191 × 95 × 48	33.3	31
2	Cow-dung	192 × 91 × 48	23.8	33
3	Cow-dung	189 × 93 × 48	34.9	37
4	Cow-dung	198 × 98 × 48	31.1	29
5	Bagasse	199 × 96 × 48	33.9	33

Source: BENS, 1996.

Table 5.7 Influence of cow-dung additive on brick quality

Cow-dung (%)	*Slop-moulded*			*Sand-moulded*		
	Strength (kg/cm²)	*Water absorption (%)*	*Density (g/cm³)*	*Strength (g/cm²)*	*Water absorption (%)*	*Density (g/cm³)*
4.8	45	25	1.37	61	24	1.45
13.0	20	30	1.22	43	28	1.39
20.0	17	33	1.22	92	29	1.31
25.9	13	34	1.15	18	33	1.21
31.0	23	31	1.19	16	34	1.20
35.5	15	35	1.15	25	25	1.31
39.4	17	35	1.12	23	33	1.07
42.9	30	31	1.27	26	31	1.21
46.0	80	36	1.13	22	32	1.19

Source: Bairiak, 1998.

Very little can be concluded from such a limited survey. There are many other variables in brick production that have an effect on the final characteristics of the product. In Practical Action's survey, it may be that one of these other variables is dominating results for compressive strength. The time–temperature conditions that the brick has been exposed to, for example, are critical in this regard. Baking a brick for 1 hour at 1,000°C produces a very different product from one baked at 100°C for 10 hours, to take an extreme example; time–temperature conditions are not a linear product. From experience, Practical Action Sudan suspect that, all other variables being equal, adding cow-dung tends to decrease the compressive strength of bricks in some direct proportion. If this is true, then the trick is to add cow-dung to the limit at which bricks with acceptable properties are still being produced.

In general, the type and quantity of additive used has a significant effect on brick quality. The Building and Roads Research Institute (BRRI) of Khartoum University has studied the influence of four organic additives on Blue Nile clay. Four sets of bricks were machine-moulded. Half of each set was sun-dried and the other half dried in shade. Once dry, both fractions were fired to the same predetermined temperature over the same time. While the mode of drying had no influence on properties, compressive strength varied greatly with the type of additive. The study found that mixing Blue Nile clay with an unspecified percentage of cow-dung (Ziballa) yielded the highest average strength, 170–180 kg/cm². Bricks that embodied groundnut shells, presumably in similar proportion, attained strengths of 120–135 kg/cm², and those with sawdust 110–125 kg/cm². Adding Garad seeds (*Acacia nilotica*) was not successful and the resultant bricks were the weakest of all sets. Note that, for all bricks, the resulting strength when fired under laboratory conditions is markedly higher for all additives than in Practical Action's field test.

Facts and figures from the brickmaking industry

Brick production is concentrated along riversides in central Sudan. During the worst of the civil war, there was almost no brick production in the south of the country. High production along the Blue Nile indicates the strength of demand for bricks in Khartoum state and the central region. Fortunately in this regard, the soil along the Blue Nile is very suitable for brickmaking. By contrast, brickmaking is non-existent along the White Nile because the soil is wholly unsuitable for that purpose. In Khartoum state and the central region brickmaking constitutes some 8 per cent of all employment. Khartoum alone absorbs about 50 per cent of the total number of workers in the national brickmaking industry.

Table 5.8 Brick production along the Nile and its tributaries

Location	*Number of production units*	*Annual brick production ('000)*	*Percentage of annual national production*
Blue Nile	1,347	2,281,280	82.40
River Nile	170	261,920	9.50
River Gash	28	39,200	1.40
River Atbara	26	36,000	1.30
Total	1,571	2,618,400	94.60

Source: Hamid, 1994.

Table 5.9 Distribution of major brickmaking centres in Sudan

Location	*Number of production units*	*Annual brick production ('000)*	*Percentage of annual national production*
Khartoum	800	1,280,000	46.2
Sennar	404	775,680	28.0
Wad Madani	60	153,600	5.5
Atbara	38	54,720	2.0
Kassala	28	39,200	1.4
Nyala	30	33,600	1.2
Al Ubayyid	20	22,400	0.8
Total	1,380	2,359,200	85.1

Source: Hamid, 1994.

Table 5.10 Employment in brickmaking industry, 1994

State/Region	*Average number of workers/unit*	*Number of production units*	*Total number of jobs*
Khartoum	22	800	17,600
Central region	21	654	13,080
Northern region	16	63	1,008
Eastern region	19	60	1,140
Darfur region	16	69	1,104
Kordofan region	16	59	944
Total		1,705	34,876

Source: Hamid, 1994.

A 1994 survey revealed in detail the quantities of raw materials - clay, additives, water and fuelwood - necessary for the production of 1,000 bricks (Hamid, 1994). The survey highlighted the well-known effect of the physical and chemical characteristics of clay on the firing energy requirement. Essentially, sandier and so more refractory soils require more energy to vitrify sufficiently for brick production. Also, the capacity of the soil for accepting the majority of additives is inversely proportional to its sandiness. At Al Ubayyid, for example, the 'clay' contains a high percentage of sand, some 67 per cent. This means that only a very small percentage of organic matter can be added before final brick quality is adversely affected, i.e. the bricks produced become crumbly and unusable. To vitrify such a refractory soil, moreover, requires a lot of energy and hence, of course, fuel. The lowest quantity of fuelwood required to fire 1,000 bricks is recorded in Khartoum and the central region i.e. along the Blue Nile. In general,

Table 5.11 Composition of clays at Gereif Shark, Khartoum

Component	*Percentage (by mass)*		
	Balabaty	*Zafota*	*Gurera*
SiO_2	44.21	49.09	75.04
Al_2O_3	17.03	15.68	8.23
Fe_2O_3	11.11	10.07	4.35
CaO	5.57	6.32	4.67
MgO	3.01	2.52	1.07
K_2O	1.75	2.53	1.26
Na_2O	0.95	2.10	0.73
LOI	12.82	8.08	5.58

Source: Practical Action/ITDG Sudan.

the survey suggests that if bricks are to vitrify sufficiently, the combined proportion of sand and additives in the mix cannot be pushed above 70 per cent.

An analysis of the clays available at Gerief Shark, Khartoum state, may be illuminating. Firstly, as most brickmakers know, a range of soil types may be found in a relatively small area. In fact, the term clay can be a misnomer as it is used to cover a range of soil types, some of which contain much more sand than clay. The figure for SiO_2 is a direct indicator of the sandiness of the soil. So, by this measure alone we would not expect to use additives *in* Gurera 'clay'. This does not of course mean that residues could not be used as auxiliary fuels by other means. Gurera soil can be expected to require a lot of firing energy; we might even question whether it is suitable for brickmaking at all. The other soils, Balabaty and Zafota, on the other hand, look good for brickmaking and have potential for the incorporation of additives. Moreover, the presence of significant fractions of known fluxes, Fe_2O_3 and CaO, which aid the process of vitrification, indicates that bricks might be fired at relatively low temperatures, thereby demanding less fuel. (Note, however, that if too much CaO is present, particularly as lumps, it can cause problems with brick quality.) One concern might be the relatively high Loss On Ignition (LOI) of these soils, particularly Balabaty. LOI is an indicator of the organic matter in the soil. This organic matter burns when bricks are fired and can cause them to exhibit low density, low strength and high water absorption.

The data for fuel consumption is given per 1,000 bricks. As we have seen in case studies from Peru, this can lead to serious misconceptions when processes

Table 5.12 Annual fuelwood consumption of brickmaking industry, 1994

State/Region	*Fuelwood per 1,000 bricks (kg)*	*Production of bricks ('000)*	*Fuelwood (tonnes)*	*%age of total*	*Wood most often used*
Khartoum	50.67	1,280,000	64,858	35.40	*Acacia nilotica* & Talh
Central region	38.70–58.00	1,172,480	49,940	27.30	*Acacia nilotica* &Talh
Northern region	66.70–150.00	90,720	9,024	4.90	Misquite, Talh, Syal, Doam, Fruit trees
Eastern region	333.30–533.30	84,000	32,900	18.00	Misquite, Shaf, Kitir, Talh
Darfur region	66.67–140.00	77,280	8,027	4.40	Talh, Kitir, Sahab
Kordofan region	66.67–400.00	64,960	18,219	10.00	Misquite, Arad, Talh, Hashab, Kitir
Total national		2,769,440	182,968	100	

Source: Hamid, 1994.

are compared. We know that brick size varies considerably from place to place, and even within one nominal location, in Sudan. Comparing 1,000 bricks from one site with 1,000 bricks from another probably means we are comparing very different masses of fired brick. That said, it is still worth noting that, from the data presented in Table 5.12, there is a huge variation in the fuelwood used to fire 1,000 bricks, i.e. from 38.7 kg to 533.3 kg. Apart from the soil type and how dry the bricks are when placed in the clamp, which we have mentioned, we must suspect that other influential factors are at work. We could speculate that, say, the condition of the firewood used might be very different from place to place. That is, the varieties listed may not only have very different essential calorific values, they may also be used in different conditions. At one location seasoned dry wood may be used, while at another small green wood with a high moisture content may be the supply. We do not, however, have sufficient data to engage in anything other than speculation, which at least raises awareness of some of the pitfalls of data collection. Such pitfalls were the reason that Practical Action developed its energy monitoring methodology.

The cost of the materials used in brickmaking represents between 45 and 64 per cent of the total production cost (Hamid, 1994). The single most expensive material input, moreover, is always fuelwood, which typically represents at least 25 per cent of production costs and can exceed 50 per cent in areas of scare supply and high demand. A 1996 study produced a breakdown of brick production costs in Khartoum (Table 5.13). Taken together, the cost of firewood

Table 5.13 Production costs for 100,000 bricks in Khartoum, 1996

Item	*Item cost (SD)*	*Cost (SD)*	*% of total cost*
A) Variable costs:		633,650	92.40
– Materials			
Animal dung	120,000		17.50
Firewood	195,000		28.40
Other	650		0.10
Total materials	315,650		46.00
– Total labour costs	318,000		46.40
B) Fixed costs		19,133	2.80
Tools			
Rent of land			
Water pump			
C) Overhead costs		33,333	4.80
Total production cost (A+B+C)		686,116	100.00

Source: BENS, 1996.

and cow-dung, which contributes as a fuel as well as a soil conditioner, make up a cost comparable with the cost of labour. A 1999 study in Kassala calculated the fuelwood cost at 44.27% of total production costs with labour and cow-dung contributing 24.59% and 7.78% respectively (Hood, 1999a).

Further potential for the use of agricultural residues and wastes

The British first introduced brick production to central Sudan, establishing the industry in the Khartoum area on the banks of the Nile and of Blue Nile. Because the soils in the area are high in clay and very plastic, the use of additives was soon introduced to minimize drying shrinkage, cracking and ultimately production losses. Although several additives, including sand, biomass materials, coal-dust and crushed broken bricks (grog), could have been selected, Sudanese brickmakers opted for cow-dung. Though it has a number of traditional uses, there was a huge surplus of cow-dung. Moreover, it was available for only the cost of transport. With time, firewood became scarce, the price soared, and brickmakers duly noted that cow-dung had another advantage over non-organic additives in that it served as a secondary fuel. From then on, this waste material attained a new value and soon had a price attached. As the price is lower than firewood, however, brickmakers tend to maximize the amount of cow-dung they incorporate in the clay mix. As we have seen, there can be an adverse effect on brick quality when cow-dung is overused.

The escalating cost of fuelwood together with technological limits on the use of cow-dung prompted research into the use of alternative wastes and agricultural residues. Sudan has vast reserves of agricultural and other residues, estimated in 2001 to amount to some 13.931 million tonnes per year (Table

Table 5.14 Characteristics of some residues

Fuel/residue	*Moisture (%)*	*Volatile matter (%)*	*Fixed carbon (%)*	*Ash (%)*	*Calorific value (MJ/kg)*
Loose bagasse	9.41	66.23	29.34	4.42	19.17
Bagasse blocks (35% molasses)	6.73	62.41	27.99	9.6	18.66
Bagasse blocks (20% clay)	?	?	?	?	14.61
Bagasse blocks (20% filter cake)	?	?	?	?	17.46
Miskit wood	14.94	76.25	21.93	1.82	19.71
Miskit roots	6.80	72.22	25.81	1.97	19.54
Cow-dung	4.02	47.93	7.30	44.77	12.81
Fuelwood (*Acacia nilotica*)	13–20	50.2	50.0	0.64	19.40

Sources: Bairiak, 1998; BENS, 1996.

5.15). The most promising of these residues for brickmaking are the stalks of various crops, peanut shells and, particularly, bagasse. In general, agricultural wastes must be processed into convenient forms for use as industrial fuel. If this is not done, high volume to weight ratios tend to make them prohibitively expensive to transport. When it comes to incorporating residues into the clay brick as an efficient way of burning them, peanut shells may need to be ground and stalks finely chopped. Though huge quantities of cow-dung are nominally available, free-range grazing of animals means that the collection of the vast majority of this waste is not economically viable. Residues not included in the table include sawdust and residues of threshing sorghum and millet (chaff).

Bagasse is a by-product of the sugar industry. It is the solid part of sugar cane that is rejected after extracting the juice. Freshly produced bagasse, which has a moisture content of around 50 per cent, is rough and coarse in texture. As it ferments naturally, however, it decomposes to soft and fine particles. In such condition it is termed 'rotted bagasse'. Heaps of bagasse are an inherent feature of the landscape around Sudanese sugar factories. Not only are these unsightly and unhygienic, encouraging rodents, the heat produced by fermentation means they tend to ignite spontaneously! Obviously, then, such heaps represent a

Table 5.15 Annual availability of residues

Residue	*Sources*	*Quantity ('000 tonnes)*
Agricultural residues	Cotton stalks	373
	Sorghum stalks	7,821
	Wheat straw	776
	Millet stalks	11,206
	Groundnut shells	1,393
	Sesame stalks	2,120
Agro-industrial residues	Bagasse	327
Animal wastes	Cattle manure (cow-dung)	55,051

Source: Second National Energy Assessment, Forestry Resources, 2000.

Table 5.16 Bagasse produced annually by sugar factories, 1994/95

	Geneid	*Sennar*	*Assalaya*	*New Halfa*	*Kennana*
Bagasse (tonnes)	200,000	300,000	215,000	268,813	750,000
%age used for steam & electricity	31.0	60	35.0	60	75
% age used for electricity (off-season)	15.7	10	12.0	11	23
Other uses (%)	9.9	4	4.8	3	2
Available surplus (%)	43.4	26	48.2	26	0

Source: BENS, 1996.

significant fire hazard. In fact, sugar factories are frequently damaged by bagasse fires. The cost and nuisance associated with monitoring bagasse heaps is a burden for the sugar factories. Nationally, there are five operational sugar factories and, over time, four of these have accumulated enormous quantities of rotted bagasse.

The use of bagasse as a fuelwood substitute is facilitated by processing it into briquettes or blocks. In Sudan, bagasse has been turned into blocks using a press designed for the production of soil blocks. Although pit molasses emerged as the most appropriate binder in trials because it is effective and low priced, it is in short supply. Other binders have therefore been tested, in particular clay and filter cake. Filter cake is a residue from sugar factories. After juice has been extracted from the cane, the bagasse is filtered through lime. The by-product product of this filtration is a mud-like cake. Using either filter cake or clay, the potential annual output of bagasse blocks from Sudan's sugar factories exceeds 400,000.

Several bagasse block production facilities were actually established at different locations. Two sizes of blocks were produced, the average mass being either 1.5 or 2.0 kg. Though these facilities operated for some time, the relatively high price of fresh molasses used as a binder, transport costs and poor marketing combined to contribute to their eventual closure. The main brickmaking locations lie considerable distances from the sugar factories and ultimately bagasse blocks were unable to compete with fuelwood. When no cost is incurred for deforestation and subsequent degradation on all scales of conception of the environment, what economists term externalities, such an outcome is predictable

Photo 5.4: Block press. Credit: Practical Action/Zul.

Table 5.17 Production of bagasse blocks at sugar factories

Enterprise/Location	*Established*	*Presses*	*Capacity (tonnes/year)*	*Production (tonnes)*	
				1994	*1995*
PDO./Assaalaya	Dec. 1994	2	500	80	NA
Muwafag/Sennar	July 1994	5	1,250	NA	NA
FNC/Guneid	July 1993	4	1,000	660	530
REC/SDC/Guneid	1995	10	2,500	290	-
Prison/New Halfa	1987	2	500	610	660
UNCHR/New Halfa	1995	10	NA	10	-
Private/New Halfa	1995	3	3,250	10	-
Total		36	9,000	1,660	1,190

Source: BENS, 1996.

and will no doubt be repeated elsewhere in the world; a false economics is thereby being employed to discredit renewable and alternative sources of energy, including wastes (Boulding, 1966; Daly, 1992; Daly, 1996; Daly, Cobb and Cobb, 1994; Ekins, 1999).

Experience of using alternative fuels

We have seen that more than 70 per cent of the energy consumed in Sudan is in the form of fuelwood, where this designation includes both firewood and charcoal. A 1994 survey of national forest products revealed that the total quantity of fuelwood consumed that year amounted to about 15.77 million cubic metres of standing wood (Hamid, 1994). Around 89.4% of this volume was consumed in the household sector, 6.8% in the industrial sector, 2.5% in the commercial and services sector and about 1.3% in Quranic schools. Firewood represents around 87.6 per cent of the total amount. The survey concluded that there was a need to conserve wood and encourage the use of alternative fuels. Although agricultural residues could play a major role in this respect, they currently contribute less than 2 per cent of national energy.

Investigating the use of bagasse as a fuel, BENS concluded that if it were compacted, it could be a viable alternative to fuelwood. It was found that applying a pressure of 48 kg/cm^2 to a mix of bagasse and an appropriate binder yields blocks of adequate strength for handling and transport. The optimum range of molasses added to bind bagasse blocks was found to be in the range of 33 to 40 per cent by mass. Other binders were also tested, filter cake and clay emerging as technically viable alternatives to molasses. With either of these binders, the recommended ratio is 20 per cent of the total mass of block. Though filter cake and clay are available at no cost, adding greater amounts to blocks has no practical advantage in terms of handling strength and it also reduces the calorific value of the block per unit mass.

The objectives of the BENS investigation then became to: 1) find the optimum level for substituting bagasse blocks for fuelwood in brick firing; and 2) investigate the effect of using raw bagasse as an alternative to cow-dung in clay mixed for making bricks. BENS carried out experiments at El Gereif Shark in Khartoum state. Two types of bricks were produced with the Blue Nile clay: bricks with cow-dung as the additive in the ratio 96 clay to 4 dung; and bricks with raw bagasse as the additive. A clamp with a capacity of 80,000 bricks was used for the experiments. Normally, such a clamp consumes around 2.7 tonnes of firewood. In addition to the additive tests, bagasse blocks were used as substitute for 50, 65 and 80 per cent of fuelwood in a series of tests. The calorific values of fuelwood and bagasse blocks were measured at, respectively, 19 and 17 MJ/kg.

Results suggest that the use of bagasse blocks as a substitute for fuelwood is technically viable and has no adverse effect on brick quality. In fact, brick quality seems to improve with the percentage of bagasse blocks used. This may be because the distribution of the blocks in the kiln gives a more favourable time–temperature firing condition, i.e. a greater percentage of blocks used as a substitute for fuelwood means that more of the fuel is in closer contact with bricks and so they are better fired. The BENS study did, in fact, conclude that the use of bagasse blocks as a substitute for fuelwood resulted in better heat distribution within the clamp. Firing bricks with fuelwood alone results in a

Table 5.18 Experimental design

Clamp no.	*Additive*	*Firewood*		*Bagasse blocks*		*Firing time (hours)*	*Cooling time (days)*
		Tonnes	*%*	*Tonnes*	*%*		
1	Dung	2.70	100	0.00	0	3.00	5.00
2	Dung	1.35	50	1.51	50	4.00	5.00
3	Dung	0.95	35	1.96	65	5.50	5.00
4	Dung	0.54	20	2.41	80	8.00	5.00
5	Bagasse	2.12	100	0.00	0	3.00	5.00

Source: BENS, 1996.

Table 5.19 Quality of bricks produced

Clamp no.	*Additive*	*% firewood*	*Brick quality category*		
			1st class (%)	*Under-fired (%)*	*Over-fired (%)*
1	Dung	100	87	10.00	3.00
2	Dung	50	90	9.00	1.00
3	Dung	35	96	0	4.00
4	Dung	20	97	2.00	1.00
5	Bagasse	100	96	1.50	2.50

Source: BENS, 1996.

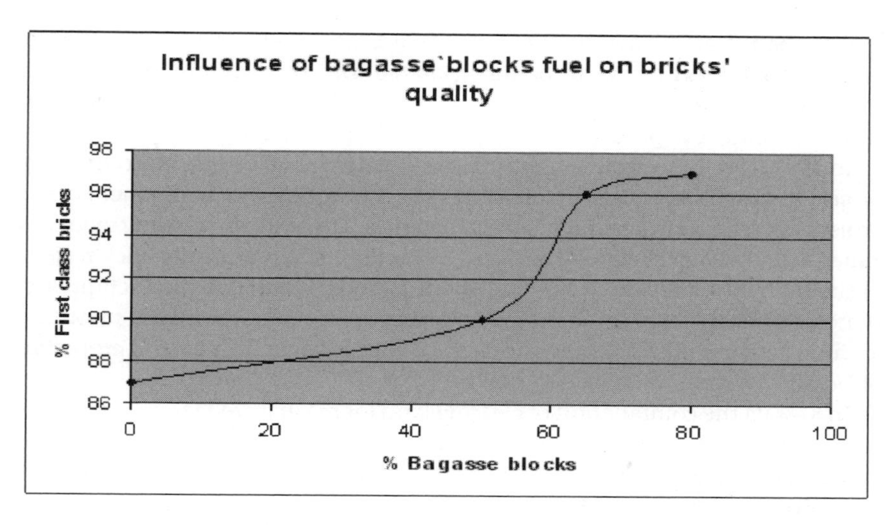

Figure 5.4 Influence of bagasse block fuel on brick quality. Contact Practical Action Sudan for further data.

Table 5.20 Physical properties of bricks produced

Clamp No.	*Additive*	*Brick dimensions (mm)*	*Water absorption (%)*	*Compressive strength (kg/cm^2)*
1	Dung	191 x 95 x 48	33.30	31
2	Dung	192 x 91 x 48	23.80	33
3	Dung	189 x 93 x 48	34.90	37
4	Dung	198 x 98 x 48	31.10	29
5	Bagasse	199 x 96 x 48	33.90	33

Source: BENS, 1996.

hot but relatively brief blaze. Consequently, energy is lost in flue gases while bricks are not maintained at temperatures commensurate with vitrification for long enough. Bagasse blocks cannot 100 per cent replace firewood, however. They are difficult to light and do not burn well in the firing tunnels. If tunnels are maintained as the technology for primary ignition of the clamp, then a proportion of fuel suitable to burning in this manner must be used.

The limiting factor on this technological innovation, the BENS investigation concluded, would be the availability of affordable and sufficient supplies of molasses. At present the cost of molasses renders bagasse block technology economically non-viable. BENS recommend further work on the use of filter cake and clay as binders in bagasse block production. With respect to using bagasse as an additive in the clay mix in place of cow-dung, no adverse affect on brick quality was noted. The calorific value of rotted bagasse (15.50 MJ/kg) is higher than that of cow-dung (13.30 MJ/kg). As a fuel substitute, therefore, proportionally less mass can be incorporated into the clay mix for a similar firing regime. Alternatively, if brick quality could be maintained, the mass of additive could be kept constant and primary fuel use reduced.

In 1997, ITDG/Practical Action Sudan implemented an integrated technology project in Kassala State, Eastern Sudan (Hood, 1999a). A major component of the project concerned shelter and building materials. The work on building materials aimed at addressing two problems faced by producers, substantively brickmakers: 1) the high cost and scarcity of fuelwood; 2) poor or variable product quality. Hence, Practical Action set about investigating the use of furnace oil as substitute for firewood in firing bricks and the effect of cow-dung on the physical properties of bricks

In Kassala the combination of clearing land for mechanized farming schemes along with other economic activities, particularly charcoal production, has led to deforestation and southward desertification. Firewood scarcity means it must be hauled over long distances from the Ethiopian border and Blue Nile province. Previously, the Miskit tree (*Prosopis*) was introduced to combat deforestation and desertification. Unfortunately, this strategy turned out to be a big mistake. Miskit depletes scarce groundwater and has encroached arable land. It has become a weed and is an enormous nuisance. The government has made it policy to eradicate Miskit nationwide. Thus, it can be cut indiscriminately for use as firewood or in charcoal burning. It can also simply be burned in situ to no productive end. Nowadays, firewood and charcoal production from Miskit provides livelihoods for tribesmen around Kassala who have lost their livestock wealth during the successive periods of drought. Competition for Miskit between the brick industry and the household sector is manifest.

Practical Action in Sudan judged the traditional brick clamp unsuitable for oil firing. In Kassala, they therefore opted to conduct trials with a small down-draught or vault kiln. In 1995, the trial kiln was duly built. Initially, however, there were problems with both the compressed air system and also the electric motor required. In addition, the kiln design had to modified. Hence, trials could not begin in earnest until the following year. Results indicated that using furnace oil was some 1.2 times more expensive than firewood. As the price of furnace oil looked set to increase and secure supply was also becoming a problem, trials were discontinued (Bairiak, 1997).

With respect to the effect of using cow-dung as an additive, results for both slop- and sand-moulded bricks indicate similar trends. As the organic matter in cow-dung is consumed during firing it leaves voids in the brick. These voids increase porosity and hence water absorption. Density, mass and compressive strength are concomitantly reduced. The absence of a Sudanese technical standard on bricks makes it impossible to propose an optimum percentage value for the use of cow-dung. In practice, though, Practcial Action in Sudan estimates that, whichever moulding process is employed, including between 20 and 30 per cent of cow-dung in the mix yields bricks of still satisfactory quality. Comparing samples with equal percentages of cow-dung added, sand-moulding

Table 5.21 Effect of adding cow-dung on brick properties

Sample	*% dung*	*Strength (N/mm²)*	*Water absorption (%)*	*Mass (kg)*
A. Slop moulding				
1	9.09	3.13	25.11	2.03
2	16.67	1.99	28.07	1.80
3	23.08	2.10	29.64	1.68
4	28.57	4.25	22.52	1.74
5	33.33	2.26	30.53	1.60
6	37.50	1.50	35.13	1.66
7	41.18	0.61	32.20	1.46
8	44.44	2.61	25.00	1.51
9	47.37	0.88	40.55	1.44
B. Sand moulding				
1	4.76	6.16	24.56	2.0
2	13.04	4.25	23.28	1.90
3	20.0	3.24	28.45	1.76
4	25.93	1.75	32.54	1.64
5	31.03	1.6	34.17	1.68
6	35.48	2.5	24.96	1.65
7	39.39	1.42	38.23	1.49
8	42.86	2.58	30.56	1.45
9	45.95	2.23	31.82	1.60

Source: Bairiak, 1997.

yielded bricks with superior properties to slop-moulding throughout the range of tests.

Practical Action Sudan instigated trials on the use of loose rotted bagasse as an alternative to cow-dung in brickmaking (Bairiak, 1998) and the use of bagasse blocks as a substitute for firewood. In a typical test, 118,000 green bricks were built into two similar sized clamps. For ignition purposes, the tunnels of both were filled with firewood (Miskit wood). The control clamp was then fired using only Miskit wood burned in the tunnels. Firing continued for 26 hours and a total of 5.83 tonnes of wood were consumed, being added at intervals of between 0.5 and 1.5 hours. After ignition with wood, the second clamp was fired with bagasse blocks for 23 hours. In this time it consumed 1.44 tonnes of wood and 3.53 tonnes of bagasse blocks. The latter were fed into tunnels at time intervals of 3 to 4 hours. The trial was controlled via sets of Buller's bars being placed into both clamps at a comparable range of locations. Because deformation actually depends on a temperature being maintained for a sufficient time, Buller's bars which sag at nominal temperatures are commonly used in the ceramics industry. In these tests, active firing of each clamp was halted when the Number

Table 5.22 Summary of clamp performances

Bricks	*Wood fired clamp*		*Bagasse block fired clamp*	
	Quantity	*%*	*Quantity*	*%*
Grade 1	35,700	64.91	42,040	66.73
Grade 2	9,000	16.36	9,520	15.11
Over-burned	2,850	5.18	3,470	5.51
Under-burned	6,000	10.91	7,000	11.11
Broken	1,450	2.64	970	1.54
Total	55,000	100.00	63,000	100.00
Firewood (tonnes)	5.83	1.44		
Bagasse (tonnes)	0.0	3.53		

Source: Bairiak, 1998.

17 bar sagged, indicating the attainment of a particular time–temperature function. By the time these bars sagged, physical indicators of complete firing were also visible: clamp height had dropped, the uppermost layer of bricks had taken on a whitish colour, emissions of smoke had all but ceased, and the insulating layer of mud had blackened and cracked. Clamps were left to cool for seven days before dismantling.

Using bagasse blocks as the principal fuel was a technical success. The percentages of Grade 1, Grade 2, over-burned and under-burned bricks produced in each clamp are very similar, while the percentage of broken bricks is somewhat lower in the clamp burning bagasse blocks. Where the technology flounders, once again, is when costs are compared. In short, the cost of transporting bagasse blocks from Halfa to Kassala, some 70 km, gives firewood a financial edge. Though it is only some SD2.62 per brick, on a 40,000 brick clamp this amounts to SD104,800 (approximately US$400). Brickmakers themselves concluded that loose bagasse was preferable to cow-dung as an additive in the clay mix. Bagasse is fine and does not have an offensive smell when wet. Dung smells and usually contains lumps that do not break down easily and leave voids in fired bricks. This effect can also mean that bricks with loose bagasse additive have smooth surface finish in contrast to that of bricks containing cow-dung, which appear pockmarked.

Having investigated options of fuel substitution and found these to be limited by financial and supply problems, Practical Action instigated trials with a different kiln technology (Hood, 1999c). The permanent, thermally insulating structure of a Scotch Kiln make it possible to achieve greater energy efficiency than when using a clamp. Furthermore, losses of under-burned bricks, which typically originate around the outer walls of the clamp, can be reduced in a Scotch Kiln for the same reason. The firing process in Scotch Kilns can also be made more controllable than it can with a clamp. Air intake can be varied and,

Table 5.23 Energy consumption and costs compared

1. Energy Additive	*Wood fired clamp*		*Bagasse block fired clamp*		
	Fuel	*Additive*	*Fuel*	*Additive*	
	Loose bagasse	*Miskit wood*	*Loose bagasse*	*Miskit wood*	*Bagasse blocks*
Calorific value (MJ/kg)	19.17	19.35	19.17	19.35	18.66
Quantity used (kg)	6,286	6,827	7,200	1,039	3,325
Energy (MJ)	120,503	132,103	138,024	20,105	62,045
Total energy (MJ)	252,606	220,174			

	Wood	Bagasse block
(A) Cost/brick including transport of fuels (SD)	11.17	13.79
(B) Cost/brick excluding transport of fuel (SD)	6.56	4.21

Source: Bairiak, 1998.

if chimneys are incorporated, dampers can be used to attain even greater control. Practical Action constructed a Scotch Kiln in Kassala, the first in the area. Members of Shambob Bricks Production Co-operative, SBPC, trained in the operation of the kiln (Hood, 1999a,b,c). Very unfortunately, due to resource constraints on the project, it was only possible to test one kiln firing with what was then the new energy monitoring methodology; the results were promising, however.

Concluding remarks

From a technological point of view, the use of bagasse, both in block form as a replacement for fuelwood and as an additive alternative to cow-dung, is a success. Environmentally too, these technologies would seem to be valid measures to counteract deforestation at the local scale, desertification at the regional scale and greenhouse gas emissions on the global scale. Unfortunately, the high cost of transport combined with no effective restriction on cutting trees for fuelwood means the technology is not financially viable at brickmaking sites that are distant from the source of supply. In areas where Miskit wood is available,

Table 5.24 Annual fuelwood consumption by traditional industries

	Brick industry	*Lime burning*	*Bakeries*	*Other industries*	*Total*
Fuelwood consumption ('000 toe)	139.32	14.04	157.38	10.4	321.14
Forest cleared annually (feddans)	2,429	243	2,738	181	5,591

Figure 5.5 Energy monitoring form, SBPC, Kassala

Name of producer: SBPC (Shambob)	Location/address: C/O IT-Kassala	Dates and time of firing: Start: 7 April 1999 at 14:00 Finish: 8 April 1999, at 18:00
Type of kiln: Scotch	Type of fuel: 1) Firewood (Miskit wood) & 2) Bagasse blocks	Mass of fuel used: Firewood: 15.62 tonnes Bagasse blocks: 3.1 tonnes
Calorific values: Firewood: 19.71 MJ/kg Moisture content 25.8% Bagasse block: 19.17 MJ/kg	Number of bricks in kiln: 96,000	Average mass per brick: Green dry: 2.11 kg Fired: 1.91 kg
Dry brick moisture content: 2.2 %	Method of forming bricks: Slop moulding	Weather conditions: Very dry and hot summer, variable wind speed
Calculation of Kiln Efficiency: Mass of green (sun dry) bricks: 202,560 kg Mass of dry bricks: 198,103 kg Total moisture content: 4,456.3 kg Drying energy: 11,546,273.3 kJ Total energy used: 287,649,688.4 kJ Firing energy: 276,103,415.1 kJ Mass of fired bricks: 183,360 kg Specific firing energy: 1.51 MJ/kg		Qualification information: Soil vitrification category: Normal < 1000°C Buller's bar No: 17 (990°C) Average firing temperature 1,000°C Firing time: 28 h

Notes: Firewood moisture content was quite variable (supply from different sources at different times) ranging between 6.53 and 37.1%; An average figure of 25.84% was used in calculations.

Comments: The quality of output bricks was quite satisfactory (more than 65% Grade 1). However, kiln firing time was very short because wood was fed to the kiln too rapidly. This led to the development of hot spots and bricks melting. Rapid firing also meant that the soaking period was short, which led to under-burned bricks at the top of the kiln. All these factors generated considerable losses (11.3%), mainly in the form of broken and over-burned bricks.

Recommendations: The firing time should be increased by reducing the firewood feeding rate, particularly on the first day of kiln firing. Also, the soaking period should be increased to a minimum of 12 hours after fire reaching the top of kiln.

Source: Hood, 1999 a,b,c

moreover, it is both government policy and an environmental necessity to reduce its prevalence.

Despite these macroscopic points, on the local level SBPC in Kassala are prospering. Through the operation of the Scotch Kiln and, when costs render it viable, replacing up to 75 per cent of their fuelwood with bagasse blocks, SBPC has increased its membership to 115 women and men and established its creditworthiness. In the year following Practical Action's intervention income increased by 20 per cent. The following year it was up 60 per cent on the initial baseline. Practical Action's thoroughgoing training involvement has helped SBPC to produce bricks that are more regular in shape and better burned, and which therefore find a good market.

One lesson Practical Action Sudan have wholly taken on board is the benefit to both parties of deeply involving producers in research. Local skills and resources are vital assets that must be nurtured and built upon rather than rejected in favour of technologies from elsewhere. That is not to say there is no scope to consider exogenous technologies, but rather that these need to be appraised and adapted to the local context. In this particular project, bagasse had been tried elsewhere in Sudan and it was possible for brickmakers to appreciate and build on the knowledge of that history which Practical Action could offer. Replacing either fuelwood or cow-dung with bagasse does not involve incomprehensible or unrecognisable technical change. Brickmakers are quick to adapt to different methods and procedures as they see how the new material performs in practice. The Scotch Kiln is, by the same token, a clear development of clamp technology that can be readily appreciated.

CHAPTER 6
Trials with boiler waste in Zimbabwe

Lasten Mika

Land and climate

Covering an area of 390,000 km, Zimbabwe is a landlocked country with an estimated population of 11.87 million. The majority of people live in rural areas but around 40 per cent live in towns and cities. To the south Zimbabwe is bordered by South Africa, to the north-west by Zambia, to the south-west by Botswana and to the north-east, east and south-east by Mozambique. Zimbabwe

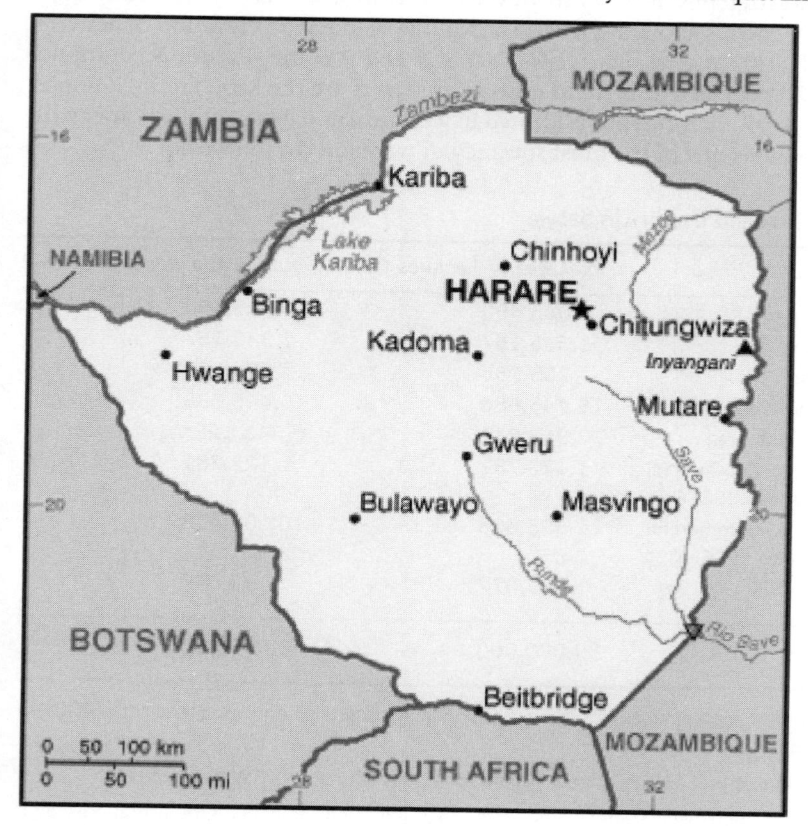

Figure 6.1 Map of Zimbabwe

is divided into 10 provinces and 57 districts and the major cities are Harare, Bulawayo, Chitungwiza, Mutare, Masvingo and Gweru.

Access to the Indian Ocean is via the ports of Beira in Mozambique and Durban in South Africa. Communication systems are generally good and the road network is fairly good with major cities and towns connected by tarred roads. The national railway system runs for about 3,400 km and connects Zimbabwe with its neighbours.

Climatic conditions are substantially influenced by altitude and there are two characteristic seasons, cool and dry winters and hot and wet summers. The winter season runs from May through July, whilst the summer season lasts from August through April. The national average minimum temperature is 15°C and the maximum is 45°C. Annual rainfall varies quite widely from region to region, but the national average is 400 mm. Rainfall is highest on the High Veld with an average annual precipitation of up to 1,020 mm. The Middle Veld, meanwhile, receives 410 to 610 mm and the Low Veld less than 400 mm.

Zimbabwe has three distinct topographical regions, namely the High, Middle and Low Velds. Dominated by a limestone ridge known as the Great Dyke, the High Veld stretches from south-west to north-west and rises to between 1,200 and 1,600 metres above sea level. The Middle Veld has an elevation of between 900 and 1,200 metres. The Low Veld is below 900 metres and accounts for around 20 per cent of the land area. Major rivers are the Sabi, Lundi, Zambezi and Limpopo. Victoria Falls is located in western Zimbabwe, on the border with Zambia, and is one of the most spectacular waterfalls in the world.

Table 6.1 Land use in Zimbabwe

Land tenure	*Total area in hectares (1996)*	*Total area in hectares (2004)*
National parks	5,045,490	5,045,490
Forest land	1,335,157	1,335,157
State land	205,255	205,255
Communal land	15,445,686	15,445,686
Resettlement area	3,958,276	5,343,126*
Small Scale Commercial Farming Area (SSCFA)	1,122,781	1,122,781
Large Scale Commercial Farming Area (LSCFA)	11,893,668	10,508,418
Town	99,077	99,077
Total	39,000,000	39,000,000

* The figures are estimates from an unpublished report by Agritex. Hence the 400 Ha discrepancy.

Sources: Ministry of Lands Annual report for 2004, Ministry of Lands, Harare.

Unpublished data base sources from Ministries and Departments: Agricultural Research and Extension (AREX)

People and nation

Zimbabwe has four main ethnic communities: Black Africans who account for the majority of the population; Whites or Europeans; Coloureds who are of mixed African and European or Indian descent; and Asians. Most Africans belong to one of two major tribal groups, the Shona who account for 71 per cent of the population and the Ndebele who constitute 16 per cent. The country's provinces are also divided along ethnic lines: Matabeleland being predominately Ndebele-speaking, Mashonaland Shona-speaking, the Midlands both Shona- and Ndebele-speaking, while Manicaland is home to a third tribal group, the Manica, and is predominately Chimanyika-speaking. Across the country a host of other minority tribal groups and languages are found, including Venda, Zulu, Sotho, Kalanga, and Tswana. Languages and cultures freely migrate across the nation-state borders that restrict the movement of people: Ndebele to the south is shared with people in South Africa and Botswana; in the east Ndau and Chimanyika are shared with people in Mozambique; and Tonga to the north crosses and re-crosses the border with Zambia. The official language nationwide in Zimbabwe is English, however. Around 45 per cent of the population are nominally Christians, while Muslims account for just 1 per cent, and the remainder follow either indigenous religions or a Syncretist faith, which is a synthesis of Christian and indigenous beliefs.

A former British colony, Zimbabwe was once ruled by a white minority who controlled the bulk of the economy, including industry, land, mines and commerce. During the 1960s the political movements Zimbabwe African National Union (ZANU) and Zimbabwe African Patriotic Union (ZAPU) launched guerrilla offensives, collectively known as Chimurenga (The Struggle), against the government. The major causes that they fought for were equitable land distribution and democracy. At the Lancaster House Conference in 1979, the warring parties agreed to the formation of a new government. On 18 April 1980 Britain recognized the country's independence as a Republic within the Commonwealth. Robert Mugabe, the ZANU leader, became Prime Minister and Rhodesia officially changed its name to Zimbabwe.

While ZANU was a mainly Shona movement and thence political party, ZAPU was principally an Ndebele organization. Political unrest in 1983 resulted in government troops cracking down on dissident activity in Matabeleland, allegedly the work of ZAPU supporters. Though the scale of the violence that followed is still contested, a great many Ndebele people were undoubtedly massacred by the largely Shona army acting on the authority of a mainly Shona government headed by a Shona Prime Minister. Some observers claim that this infamous military operation, known in Ndebele as Gukurahundi, amounted to ethnic cleansing or attempted genocide. Certainly, it is a terrible landmark in Zimbabwe's post-colonial history. Military action and opposition to it effectively ended with the signing of a peace agreement between ZANU and ZAPU in 1988.

In December 1990 legislation was passed that allowed for the confiscation of White-owned farms with compensation to be determined by the government.

To cut a very long, repressive and bloody chapter short, the country is currently characterized by bad governance and absolute official intolerance of opposition politics and any form of dissent. The government largely blames the West for the catastrophic political and economic situation. President Mugabe is particularly antagonistic towards Britain, which he accuses of attempting to re-colonize the country. Meanwhile, both the European Union and the United States have imposed economic sanctions on Zimbabwe, citing widespread and gross human-rights abuses.

Society and the economy

In the first decade after independence, poverty reduction was a priority for the government and the social sector received the lion's share of the budget. This resulted in the growth of education, health and agriculture. Through the 1990s, however, a combination of poor governance, corruption, inappropriate economic reforms (not least the IMF-imposed Economic Structural Adjustment Programme, ESAP, referred to by Zimbabweans as Even Sadza's A Problem, sadza being the nation's staple food derived from maize), and natural disasters turned back the clock on the nation's social achievements.

The Zimbabwean economy since independence is a tale of two contrasting periods. In 1980 the country inherited a dual economy characterized by a relatively well developed modern sector and a largely poor rural sector that employed 80 per cent of the labour force. This imbalance was addressed by the government through a number of plans and strategies such as Growth with Equity (1981), Transitional Development Plan (1982–85) and the First Five Year National Development Plan (1986–90). According to the *Zimbabwe Millennium Goals Development Report* (2004), policies were aimed at poverty reduction,

Table 6.2 Selected demographic and social indicators

Population (2002)	11.87 million
Population growth rate	1.1%
Urban population as %age of total population (2002)	39.8%
Population under 15 years of age (2002)	40%
Population 65+ years of age (2002)	3%
Average life expectancy (2005)	37 years
HIV/AIDS prevalence rate (2005)	21.3%
Infant mortality rate per 1,000 live births (2004)	78
Average household size (1999)	6.4 people
Literacy rate (age 15–24 years)	97.6%
Unemployment rate (2005)	>50%

Sources: Zimbabwe Millennium Development Goals Progress Report, 2004; Preliminary Census Report 2002; UNAIDS EPI Update 2005.

improving rural infrastructure and addressing inequalities, particularly via a consensual programme of land reform.

The period 1980 to 1990 is marked by rapid gains in social and economic development. By contrast, the period from 1990 onwards is characterized by widespread and substantial erosion of all the progress achieved. Zimbabwe's catalogue of economic woes makes grim reading: plummeting GDP, upward spiralling unemployment and under-employment, hyperinflation, rapidly declining agricultural productivity, and the failure to implement economic reforms. Real GDP growth rate between 1991 and 1995 averaged 1.5 per cent per annum against a population growth of 2.2 per cent. Hence, there was no meaningful poverty reduction and no employment creation. Against this background, extreme poverty increased significantly during the 1990s. An estimated 35 per cent of households lived below the poverty line in 1995, up from 26 per cent in 1990.

Since 1999, Zimbabwe's economy has been ranked as one of the world's fastest shrinking. Overall, it is estimated that the economy has contracted by around 40 per cent. In 2005, shortages of basic commodities and key inputs, such as fuel and electricity, characterize the situation. These shortages undermine production and increase production costs. The unemployment rate in the formal sector rose to above 70 per cent, while the informal sector ballooned. Poverty levels rose sharply, with 80 per cent of the population living below the 'Total Consumption Poverty Line', according to an UNDP-MDG report. The UNDP ranks Zimbabwe 145th in an assessment of 175 countries.

Home-grown initiatives to salvage the economy, such as the Zimbabwe Programme for Economic and Social Transformation (ZIMPREST) in 1998, the Millennium Economic Recovery Programme (MREP) in 2001, and National Economic Revival Programme (NREP) in 2003, have all been in vain. Bad governance, resulting in the withdrawal of most of the donor community, international isolation, and a lack of resources, have contrived with recurring

Table 6.3 Key economic Indicators

Indicators	*1990*	*1995*	*2000*	*2002*	*2003*	*2004*
GDP growth (%)	7.0	0.2	–8.2	–14.5	–13.9	–2.5
Per capita GDP growth (%)	5.5	–1.3	–7.7	–14.7	–14.1	–
Inflation (%)	15.5	22.6	55.9	133.2	622.4	123.7
Overseas development aid (US$ million)	295.9	347.7	192.6	–	–	–
Net foreign investment (US$ million)	–12	98	16	–0.3	22.6	3.5

Source: Zimbabwe Millennium Goals Development Report, 2004 *(draft).*

droughts and floods to undermine efforts at recovery. In curt summary, virtual economic collapse means Zimbabwe currently faces total societal breakdown while the spectre of mass famine looms large.

Environment matters

The major environmental problems facing Zimbabwe are water supply, loss of biodiversity, land degradation and soil erosion. In part at least, these problems are associated with Zimbabwe's colonial past. The Rhodesian era was characterized by social engineering that saw black Africans not needed as industrial labour or domestic servants confined to designated communal lands. These areas were most often marginal lands, not required by the white commercial farmers. Environmentally vulnerable communal lands became overpopulated and under-resourced, resulting in loss of biodiversity, deforestation, land degradation, soil erosion and, in some areas, desertification. In the communal lands, poor farming practices, such as overgrazing and excessive tree felling, exacerbate the natural vulnerability associated with 'thin' or shallow topsoils and slow rates of soil formation. Due mainly to human activity, some 500 million tonnes of soil are reckoned to be displaced annually. This translates to an estimated 50 to 75 tonnes per hectare of soil loss.

In the upheavals through the 1990s and into the new millennium, settlers on land the government has acquired by one means or another have had a significant environmental impact. They have cleared land for agriculture, to obtain wood for building houses, and for domestic fuelwood. Patently, this is exacerbating deforestation and land degradation nationally. While exact figures are unknown, estimates suggest that deforestation ranges between 100,000 and 320,000 hectares per year. This impact is not, however, solely down to land reform and resettlement. In some parts of the country, particularly the eastern highlands, commercial forest plantations are a major industry. In an attempt to meet wood supply needs, the introduction of exotic species, such as eucalyptus, has put further pressure on indigenous species. These exotic trees tend to be fast-growing and demand a lot of water. This leads to a decline in the water table and groundwater supplies, which most affects the less aggressive indigenous species.

Nationwide, there are a number of other environmental impacts worthy of consideration. Droughts and floods degrade cultivated land and rangelands in some parts of the country. This has an impact on plant cover, livestock numbers, and consequently agricultural productivity. In national parks, meanwhile, food shortages have led to an increase in poaching and the decimation of certain species of wildlife. These days, not only are elephants illegally shot for their tusks, but all manner of smaller mammals and birds are shot for meat. Elsewhere, mining operations, including small-scale gold and diamond mining, are also insensitive to environmental impacts. Effluent from mining waste dumps has polluted surface water and the sector has also been responsible for the build-up of silt in many streams and rivers. Commercial mining has, in addition, had an impact on the social environment, displacing communities from the ore-rich areas where companies wish to operate. Zimbabwe's main source of carbon emissions is coal burning. Apart from in Harare, air pollution is not considered a serious problem. When there is the fuel to run them, vehicle emissions are a significant contributor to both air pollution and carbon emissions.

The Environmental Management Act (Chapter 20: 27 No 13/2002) of 2004 provides a framework for mainstreaming environment into national policies and programmes. The challenge remains to build capacity at both the national and local levels to ensure effective implementation of the Act as well as link EMA with other legal instruments, such as the Traditional Leaders Act, to make environmental management more effective. Other institutional initiatives that should provide support for operationalizing the principal of sustainable development include:

- consultative and planning forums of the Convention to Combat Drought and Desertification;
- District Environmental Action Plan (DEAP);
- Communal Area Management Programme For Indigenous Resources (CAMPFIRE);
- Water Act;
- Rural Electrification Programme;
- urban and peri-urban councils.

The energy situation

The main sources of energy in Zimbabwe are coal, wood, electricity and petroleum fuels. Fuelwood is estimated at providing the bulk (53%) of the total supply, followed by coal (20%), liquid fuels (14%) and electricity (13%). The energy supply situation is characterized by liquid fuel shortages and intermittent power cuts.

According to the Zimbabwe Power Company, the nation's peak electricity demand is projected to have increased from 2,000 MW in 2004 to over 2,600 MW by 2007. Access to electricity is estimated nationally at around 40 per cent on average, but access in the rural areas is much lower at about 19 per cent. The

Table 6.4 Percentage of households having access to energy sources in Zimbabwe

	Energy for cooking					
	Urban areas		*Rural areas*		*National*	
	Poor	*Non-poor*	*Poor*	*Non-poor*	*Poor*	*Non-poor*
Electricity	73.1	81.9	2.1	11.0	19.0	52.8
Kerosene	39.7	33.7	1.0	13.5	10.2	25.4
Wood or Coal	12.7	5.4	98.6	80.6	78.1	36.3

Source: Central Statistical Office, Zimbabwe.

country currently imports more than 40 per cent of its electricity from neighbouring countries. A shortage of hard currency has meant Zimbabwe defaulting on payments for electricity imported from South Africa, Zambia, Mozambique and the Democratic Republic of Congo (DRC). The resultant power shortage has forced the government to introduce load-shedding, i.e. scheduled power cuts. Load-shedding has impacted negatively on the productive and service sectors, threatening their viability at a time when the economy needs them most. In the long term, in any event, Zimbabwe cannot rely on meeting its needs by importing regionally. By 2007, forecasts indicate that the other countries in the region may not even be able to meet their own needs. Meanwhile, lack of investment and the economic crisis generally means that Zimbabwe is largely unable to maintain or renew its aged power stations and electrical infrastructure. So desperate is the economic situation that thieves have taken to stealing the cooling oil from substation transformers to use in vehicle engines.

The traditional biomass energy sector has continued to play an important role in the energy economy of Zimbabwe. Seventy five percent of the rural population meet 80–90 per cent of their energy requirements from traditional fuels, principally wood. Almost all rural households use wood as their main source of fuel for cooking. In urban areas, as the economic crisis bites ever deeper, wood has re-emerged as a key energy resource for both the poor and also the affluent. This is the direct consequence of shortages of kerosene and LPG along with intermittent load-shedding and a lack of access to electricity, especially in peri-urban settlements. Small enterprises and industry are also significant users of fuelwood. In the informal food industry, wood is the main source of energy with LPG and electricity used only to a very limited extent. In traditional brickmaking, wood is far-and-away the most significant fuel. The main fuel used to cure tobacco, formerly an export mainstay of the economy, is also, predictably, wood.

Annual deforestation in Zimbabwe is estimated at 1.5 per cent of all woodland areas. Principal causes are the clearing of land for agriculture, the extraction of fuelwood for domestic and agro-industrial purposes, timber felling for construction, and forest fires, particularly in times of drought in susceptible areas. Considered on the national scale, there is not yet a shortage of fuelwood. In some rural 'communal areas', however, shortages can be acute. Obviously, these shortages are related to population density and the time period over which the area has been settled.

Zimbabwe does not have known oil reserves. The transport sector relies totally on imported liquid fuels brought in by pipeline from Beira in Mozambique. Kerosene is used mainly in the household sector, but also to a limited extent by industry. The effect of the shortage of foreign exchange has been manifest most dramatically in the petroleum fuels subsector. From about 2000, petroleum fuel supplies have been erratic. Long queues of motorists at fuel outlets have been all too frequent, and images of these epitomize Zimbabwe's economic plight in the international consciousness.

Coal is a major energy resource for Zimbabwe, with extensive reserves to be found at Hwange in particular. Coal provides the bulk of industrial energy and also fuels power stations to produce about 70 per cent of the nation's electricity. Large-scale commercial farmers use coal for curing tobacco. In medium- to large-scale industry, coal is used extensively to fire clay bricks. There is scope for increased coal use in small-scale processing industries such as brickmaking, food preparation and manufacturing. There is little household consumption of coal: the high sulphur content of the indigenous resource makes it particularly unsuitable for use as a fuel for domestic cooking on open fires.

The housing situation

Traditional housing in most of Zimbabwe was built with pole and *dagga*. This material matrix consists of mud and cow-dung plaster spread over a framework of upright poles and interwoven saplings. The roofs of traditional houses are roughly thatched with straw. Brick construction was introduced well before independence, however, and is an established building technology. Even in rural areas, bricks are, or would be, the walling material of choice for many people. The demand for bricks in urban areas has long outstripped supply because there is a critical shortage of housing. In 1985 it was estimated that the urban housing backlog stood at over 1 million units. To alleviate this backlog, the government targeted constructing 162,500 units per annum from 1985 to 2000. Actual annual construction was only in the region of between 15,000 and 20,000 units, however. Since 2000, the construction rate has declined still further. In 2002, for example, only 5,500 house stands were serviced, i.e. provided with water and sewage connections and electricity. According to the *Zimbabwe Millennium Development Goals 2004 progress report*, a quarter of a million houses need to be constructed and serviced annually.

Homelessness is widespread across the country, especially in urban areas. The result is squatter camps or illegal settlements, massive overcrowding in high density suburbs where the majority of poor urban people live, and the proliferation of illegal backyard shacks – 'tangwenas' - that do not have any direct services. This situation has strained the capacity of both central and local government to breaking point. They simply cannot cope with the demands on the infrastructure of towns and cities. The number of people deemed to be living in slums is estimated at 157,000. In 2005, a programme of demolishing dwellings that the government judges illegal has exacerbated the problem. Politically motivated, the programme was an attack on urban supporters of the main opposition party, the Movement for Democratic Change (MDC). People made homeless, often having been forced to demolish their own homes, have been directed by government to return to their traditional homes in the communal lands. This does not approach any sort of solution to the housing crisis, however. Communal lands do not have the resources to house or employ the influx of refugees, which was why many people migrated to cities in search of work in the first place.

> The government's urban slum demolition drive in 2005 drew more international condemnation. The president said it was an effort to boost law and order and development; critics accused him of destroying slums housing opposition supporters. Either way, the razing of 'illegal structures' left some 700,000 people without jobs or homes, according to UN estimates. (BBC, 2005)

Brick production

The fired brick industry has been dominated by large-scale plants. Bricks are heavy and so the cost of transport plays a part in the location of these plants. Hence, most are located in areas with suitable soils that are fairly near urban centres of demand. The early 1990s witnessed the emergence of some medium- and small-scale producers, complementing rather than competing with large-scale plants. Demand for bricks in 1991–92 stood at 800 million bricks while annual production was only 350 million. At that point in time, moreover, demand was increasing at the rate of 15 per cent annually. The shortfall in production obviously meant a shortfall in supply and the price of bricks increased dramatically, attracting the interest of entrepreneurs on a range of scales. One of the barriers to market penetration by small-scale producers has been the punitively high standards put in place before independence. These are based on British standards and are inappropriate for many indigenous construction needs. In 1992 the Mugabe government finally relaxed the standards applying to building materials that can be used in urban areas. This opened up a market for so-called 'farm bricks' from small-scale peri-urban producers.

The brickmaking industry in Zimbabwe can be broadly classified into three scales of enterprise. Large-scale plants are categorized as those that produce, or could produce, over 30,000 bricks per day. Production of bricks on this scale is quite mechanized, employing mechanical diggers in the excavation process, mechanized crushing and size screening, extruding plant, and conveyor belts and motorized vehicles in the handling process. Manual labour is typically restricted to loading and offloading kilns. Most large-scale producers use beehive kilns, though some employ Hoffman Kilns, tunnel kilns and even large clamps. The most common fuels used are coal and coal-dust. Major brickworks are situated near the capital, Harare, and also near the largest city in Matebeleland, Bulawayo.

Medium-scale plants are considered to be those that produce between 10,000 and 30,000 bricks per day. Generally, the equipment used is similar to that used by large-scale producers, the differences being the age and condition. This scale of enterprise usually relies on second-hand equipment and is therefore prone to frequent breakdowns that obviously have an adverse effect on productivity. Clamps or scove kilns (i.e. a brick clamp that has firing tunnels built into it) are the principal firing technology because they require no capital investment in infrastructure. Once again, the main fuels used are coal and a limited quantity of

Photo 6.1 Zimbabwean beehive kiln. Credit: Lasten Mika.

coal-dust. Production is continuous throughout the year, with green bricks and unfired clamps being covered with plastic sheets during the rainy season.

Small-scale brickmaking enterprises are those that produce less that 10,000 bricks per day. Operations are characterized by being labour-intensive from excavation through to offloading of the clamps. A typical clamp would contain about 30,000 bricks, but they range in capacity from only perhaps 5,000 to 50,000 bricks. Some mechanization can be observed in the moulding process where a few enterprises have introduced manual presses. In the main, though, the traditional slop moulding method is employed. The moulds used vary in size, resulting in bricks of somewhat variable dimensions being produced in different parts of the country. Typically, bricks are fired in scove kilns. The main source of fuel is wood, though, in a limited number of cases, coal, coal-dust and boiler waste are employed. The number of people involved in the production of bricks in each enterprise varies from just two up to perhaps ten workers. Brickmaking is seasonal with production typically suspended during the rainy season, a long lay-off that for many brickmakers may well last from November to March.

'Farm bricks' is the generic term used to describe the type of brick produced by small-scale enterprises in Zimbabwe. Brickmakers obviously choose their working sites because the clay available there is suitable to purpose. With very small-scale rural producers, the site chosen is often close to an anthill. Soil from anthills is widely acknowledged as good for brickmaking. Evidently ants choose

Table 6.5 Relative brick quality and energy requirement

	Large-scale	*Medium-scale*	*Small-scale*
Avg. density green bricks (kg/m³)	1,775	1,760	1,600
Avg. density fired bricks (kg/m³)	1,700	1,600	1,550
Water absorption by mass (%)	14.00	16.50	17.00
Avg. compressive strength (MPa)	14.00	13.50	6.00
Relative energy requirement	2.00	1.04	1.00

Source: Practical Action Zimbabwe.

soils that tend to have the right proportions of sand and clay because that is what is required to build their homes. Moreover, the soil in anthills is homogenized, free from lumps and stones. Traditionally, testing the suitability of soils for brickmaking is a process of trial and error whereby a few sample bricks are moulded and baked in an open fire. At some sites, in rare instances, it is necessary to modify the clay/sand balance of the soil. In general, this is only worthwhile when soils contain too much clay but there is an available source of river or pit sand not too far away.

Before moving on to Practical Action's project intervention, it is worth considering the processes involved in small-scale production in a little more detail. Brickmakers excavate soil using simple hand-tools, picks and shovels. The clay then is transported in wheelbarrows to the mixing place where water is added. There is no crushing, sieving or tempering of the soil. Moulding

Photo 6.2 Excavation and mixing of clay. Credit: Practical Action/Zul.

Photo 6.3: Building a brick clamp. Credit: Janet Boston

typically involves placing a wet mass of clay in a wooden mould with three brick compartments. The moulder stands in a pit up to the waist and moulds bricks on the ground. Excess clay is removed from the top of the mould with a wooden scraper. A labourer then carries the mould to a clean and level drying area where the bricks are demoulded. Wet bricks are spaced to allow for air circulation and are left to dry for between three and five days, depending on weather conditions. The half-dry or green bricks may then be stacked to dry more fully. Traditional brickmakers may, however, begin to build the half-dry bricks into a scove kiln after only a few days of drying. The scove kiln has firing tunnels built into its base and when complete is 'scoved', i.e. plastered with soil and water mixture. Sometimes brick rubble from the breakages of previous firings is used to cover the top of the scove kiln. The kiln is then fired with wood burned in the tunnels. Ideally, a slow fire is maintained for some time to fully dry the bricks. The intensity of the blaze is then increased and maintained for a period. Finally, brickmakers stop feeding wood to the fire and the kiln is allowed to cool naturally before being dismantled, typically after a week or more.

The environmental impacts of brickmaking include the land and landscape degradation that can result from clay extraction, local air pollution and the emission of carbon dioxide. A concern with small-scale brickmaking is the open pits left when brickmakers vacate a site. These are a danger to people and animals. Moreover, small-scale brickmaking certainly contributes to deforestation, which is particularly critical in the fragile ecosystems where such operations tend to be sited. As Table 6.6 indicates, the contribution to air pollution and carbon dioxide by small-scale brickmakers appears negligible when compared nationally to that of large- and medium-scale producers. Local environmental impacts are perceived as most severe where there is a high concentration of production units on the fringes of an urban area.

The intervention of Practical Action

One impact of IMF-imposed economic structural adjustment programmes in the early 1990s was to increase unemployment, with many jobs cuts occurring in Zimbabwe's still extensive civil service. This forced many people to seek alternative work outside the formal sector. Given the shortage of bricks and the opportunity for income generation, brick production was one sector that attracted a number of people, both entrepreneurs and labourers. New and augmented small-scale production facilities encountered a number of obstacles,

Table 6.6 Environmental impacts of the brickmaking industry

	Large-scale	*Medium-scale*	*Small-scale*
Emission of carbon dioxide per annum (kg x 10^6)	3.5	1.25	<0.25
Sulphur dioxide emissions per annum (kg)	50,000	11,000	<2,000

Box 6.1 Assessment of brickmaking at Price Busters

An assessment was carried out at Price Busters Brick Company in 1996. The assessment was made to determine the pollution levels and the energy efficiency of the kilns and other processes. Emission levels measured included carbon monoxide (CO), carbon dioxide (CO_2), sulphur dioxide (SO_2), nitrogen oxides (NO_x) and particulate emission (PM-10).

The company employed 450 workers who worked two shifts per day. It has two extrusion machines each producing 60,000 bricks per shift. Three types of burning kilns are used: beehive, Hoffman and field clamps. Coal is the energy source.

On-site observations showed that the raw materials were not well defined and the lack of specifications was contributing to high losses. The extrusion machines were too old (over 50 years) affecting productivity due to frequent breakdowns.

Test results

Beehive kiln	*Measured value*	*Standard*
PM-10	87 mg/m^3	115 mg/m^3 (Zimbabwe)
SO_2	0.422 ppm	0.03 ppm (NAAQS)
CO	230 ppm	9.0 ppm (NAAQS)
CO_2	2215 ppm	35 ppm (NAAQS)
NO_x	0.153 ppm	0.053 ppm (NAAQS)
Field clamp	*Measured value*	*Standard*
PM-10	87 mg/m^3	115 mg/m^3 (Zimbabwe)
SO_2	0.51 ppm	0.03 ppm (NAAQS)
CO	180 ppm	9.0 ppm (NAAQS)
CO_2	1840 ppm	35 ppm (NAAQS)
NO_x	0.215 ppm	0.053 ppm (NAAQS)

- Efficiency of coal combustion was quite high
- Dust emissions higher than level permitted by National Ambient Air Quality Standards (NAAQS)
- Gas pollutants were higher than allowed limits

however. Inappropriate standards meant that, if they wished to supply the potentially lucrative urban market, these enterprises were obliged to produce bricks that could be classified as 'commons'. In general, producing bricks with the compressive strength and water absorption demanded by the commons standard is not beyond small-scale brickmakers. Production of commons does mean a step change in technology, however, particularly with respect to quality control.

At the time, interest rates were high, discouraging brickmakers from taking loans to make capital investments in, for instance, moulding machines or kilns. Intermediate technologies were potentially an appropriate means of production for small-scale producers. Characterized in part by relatively low investment costs, such technologies presented the possibility of making the step-up from farm to common brick production. Recognition of this potential signalled the conceptualization of a Practical Action building materials project in Zimbabwe. The project focused on three main areas of small-scale brick production:

- increasing brick quality;
- introducing alternative fuels;
- improving firing techniques.

The economic rationale for improving brick quality at the time has been outlined already. Practical Action staff also recognized from the project's inception that increased brick production on the scale required nationally would mean an untenable burden on forests and woodlands. If wood continued to be the main fuel used by the small-scale sector, the predicted increase in demand could critically exacerbate the already acknowledged environmental problem of deforestation, subsequent land degradation and soil erosion. Domestic demand for fuelwood meant that these problems would tend to be most acute in exactly the areas where brickmakers worked. Thus, the project focused on identifying alternative fuels. Much the same reasoning vis-à-vis deforestation, gave rise to the focus on improving firing techniques, i.e. increasing energy efficiency as well as ensuring bricks were sufficiently fired to meet the quality standard. Traditional small-scale brickmaking technology tends to be very inefficient and thus energy-intensive and damaging to the environment, specifically in terms of deforestation and carbon dioxide emissions.

The project was implemented both via training centres and through existing brickmaking enterprises. At the training centres, improved brickmaking was integrated into construction-related courses. This was a deliberate strategy designed to ensure that graduating students disseminated the technology when they returned to their home areas. A number of training centres across the country became active project partners: Fambidzanayi (Harare), Mupfure (Chegutu), Kaguvi (Gweru), AVOCA (Plumtree), Hlekweni (Bulawayo) and Chipiku (Bikita). The other strand of the project was work with existing brickmaking cooperatives, notably Kuwirirana, located in a high-density suburb of Harare, and Kurehwasekwa, situated in Epworth, a largely illegal settlement on the fringe of Harare. The objective of this strand of the project was to pilot small-scale brickworks capable of supplying the urban market and doing so as viable businesses, capable of record-keeping and ultimately formalizing their operations. In other words, enterprise development was as important an element of the work as technology change.

As a first step, studies were made with small-scale brick producers to identify the causes of inferior brick quality. One conclusion was that the method of moulding needed to change in order to meet the required common standard. Slop-moulding made it extremely difficult to reach the required compressive strength, water absorption and dimensional accuracy. Manually operated brick-moulding machines were therefore imported for trials, with a long-term view to their local manufacture. The first institution to adopt such technology was Mupfure College, located south of Harare. The first machine extensively tested there was a press made by IT Workshops in the UK. It is designed for moulding to be performed at a comfortable working height. A clot of prepared soil is thrown into a chamber lubricated to facilitate release of the moulded brick. Closing the chamber lid and giving the clay a forming press via the mechanism of a foot pedal ensures the brick is dimensionally accurate with sharp edges and corners.

It is good soil preparation prior to moulding and effective firing that exert most influence on the brick's compressive strength and water absorption, however. The results of trials at Mupfure led Practical Action to design, make and test a table for sand-moulding that had no moving parts to wear out. Once appropriate local materials had been identified, this moulding table design proved durable and affordable. Meanwhile, a second moulding machine was imported for test. The Belgian Ceratec press was actually designed for the production of soil blocks, including stabilized soil bricks, SSBs, which are soil-stabilized with a binder, usually lime or cement. Unlike the IT Workshops machine, then, the Ceratec is a bona fide press. Consequently, the soil mixed used to make bricks has to be quite dry; it is not possible to press water. The Ceratec press yielded very good-looking and dimensionally accurate bricks. However, unless the soil used had been thoroughly mixed and moistened beforehand, as well as allowed to dry sufficiently before pressing, there was a tendency for the final product to be rather brittle.

When the project turned its attention to alternative fuels, fuelwood was so scarce around many urban centres that brickmakers either had to buy wood brought in from rural areas or steal from nearby farms and managed woodland. Typically fuelwood from rural areas would travel more than 200 km. Practical Action's first initiative was to pilot the coal-fired clamp as an alternative to the scove kiln in areas, specifically around Harare, where coal was available. Rather than being fired via tunnels at its base, the coal-fired clamp has coal laid between layers of bricks. Otherwise the clamp shape, capacity, building technique and scoving were familiar to brickmakers, so the technology change was manageable. Once the bottom layer of coal is ignited utilizing a temporary external grate, clamp-firing proceeds automatically, though air flow can be controlled somewhat and the clamp can be shielded from winds.

This technology was quickly adopted by Practical Action's cooperative partners and thence, via a programme of peer-to-peer visits and training, to a good number of peri-urban brickmakers nationwide. The main problem with the coal-fired clamp was not technological, in fact, but financial. Dealing with formal coal merchants meant brickmakers had to pay for supplies in advance of firing. Previously, they had, in many instances, been able to pay informal sector suppliers of fuelwood after selling the bricks produced. Though many small-scale enterprises could manage this cash-flow problem, it remained the case that coal was expensive, offering at best only a marginal saving over fuelwood. The main benefit of coal was its relatively constant availability rather than its economic viability. So, Practical Action began to look for other alternative fuels. The huge quantities of boiler ash from Harare's thermal power station presented the management there with a disposal problem. In fact, the ash was free of charge to anyone who wanted to take it away. Luckily for brickmakers, if not for society and the environment in general, the power station is old and inefficient. The boiler ash therefore retains a percentage of unburned carbon and so a calorific value. Practical Action's partners first experimented successfully with using the ash instead of a proportion of coal in the clamp. Thereafter, they also had some success with incorporating boiler ash into the body of bricks.

Table 6.7 Typical calorific values of fuel in Zimbabwe

Fuel type	Calorific value (kJ/kg)
Wood	15,000–18,000
Coal	30,000–32,000
Coal boiler waste	10,000–15,000

In 2005, it was common for peri-urban brickmakers to use both coal and also coal boiler waste when these fuels were available. Apart from thermal power stations, boiler waste is available from tobacco farms and other industries that utilize steam. Unfortunately, suppliers have recognized its value and boiler ash is no longer freely available. It is, though, the cost of transport, and indeed the availability of fuel for transport, that discourages the use of coal.

Practical Action's work in monitoring and thence increasing energy efficiency was cut short by the combination of the crisis in the country and a lack of project funding. A first series of tests using the energy monitoring methodology that was so successful in Peru yielded the results set out in Table 6.8. Fieldworkers recorded: 'There was a large variation in the results obtained

Table 6.8 Results of energy monitoring

Parameter	*Epworth Kiln 1*	*Epworth Kiln 2*	*Epworth Kiln 3*	*Epworth Kiln 4*	*Epworth Kiln 5*
Avg. mass green brick (kg)	3.56	3.60	3.28	4.84	3.55
Avg. mass fired brick (kg)	3.33	3.40	2.93	4.63	3.30
Moisture content green bricks (%)	6.91	7.00	0.50	4.54	0.50
Type of fuel	Eucalyptus	Eucalyptus & boiler waste	Boiler waste	Mix of hard woods	Boiler waste
Mass of fuel (kg)	3,580	1,693 & 1,595	5,093	3,005	5,093
No. of bricks	20,000	20,000	23,245	8,000	18,000
Specific firing energy (MJ/kg)	0.47?	0.50?	0.65?	1.04?	1.17?

Table 6.9 Production cost comparison (Zimbabwe dollars, Z$)

	Wood	*Wood & boiler waste*	*Boiler waste*	*Coal*
Clamp parameters				
Monthly production rate	40,000	40,000	40,000	40,000
No. of clamps fired per month	1	1	1	1
Bricks per month	20,000	20,000	30,000	25,000
Breakage rate per clamp	12	8	2	2
No of bricks available for sale per month	17,600	18,400	29,400	24,500
Fuel input				
Amount of fuel (wood) per clamp	3,580	1,693	–	–
Amount of fuel (waste) per clamp	–	1,597	5,093	–
Amount of fuel (coal) per clamp	–	–	–	2,000
Cost of fuel (wood)	8,950,000	4,232,500	–	–
Cost of fuel (waste)	–	343,200	1,092,000	–
Cost of fuel (coal)	–	–	–	15,500,000
Transport cost (wood)	1,200,000	1,200,000		
Transport cost (waste)		1,200,000	1,200,000	
Transport cost (coal)				1,200,000
Total cost of fuel	**10,150,000**	**6,975,700**	**2,292,000**	**16,700,000**
Labour input				
Cost of 5 people per month	4,000,000	4,000,000	4,000,000	4,000,000
Subtotal costs (direct Inputs)	14,150,000	10,975,700	6,292,000	20,700,000
Fuel as percentage of direct costs	72%	64%	36%	81%
Overheads				
Estimated at 10% of subtotal costs	1,415,000	1,097,570	629,200	2,070,000
Summary				
Cost of brick production	15,565,000	12,073,270	6,921,200	22,770,000
Unit production cost	884	656	235	929
Selling price per brick	1000	1000	1000	1000
Profit per brick	116	344	765	71
Profit margin	**13%**	**52%**	**325%**	**8%**

Source: Practical Action Zimbabwe.

for specific firing energy... This could not give conclusive results and thus further investigations are recommended.' The results for specific firing energy attained in the Epworth tests were so low that they were instantly suspect. For one thing, fieldworkers did not systematically record how well-fired the bricks from the five trial kilns were. Hence, they could make no relative judgement about whether, for example, bricks from Kiln 1 were under-fired compared to Kiln 5. The tests indicated that the builders and operators of the kilns had a major influence on the specific firing energy required. Moreover, fieldworkers suspected significant errors in the data recorded by brickmakers, particularly with respect to the masses of fuels used. Unfortunately, for the reasons given, there was no further opportunity for investigation or for fieldworkers to familiarize themselves with the methodology.

In 2005, Practical Action in Zimbabwe performed a rapid appraisal of the financial implications of using different fuels in brickmaking. The results are recorded in Table 6.9. The indication is that profit margin will vary considerably according to fuel choice. Analysis suggests that wood, a small volume of which was purchased at commercial rates in Harare in 2005, is the most expensive fuel option and boiler waste the cheapest.

Practical Action in Zimbabwe report that a nationwide survey conducted in 2005 shows there is a high level of awareness among brickmakers about the use of coal and boiler waste as alternatives to fuelwood. Many brickmakers were familiar with both coal-fired clamp technology and also the potential for integrating a proportion of the fine fraction of boiler waste as fuel into the body of the brick. As expected, the survey indicated that the factors determining which fuel is used are availability, the cost of fuel, and the cost of transporting it. Practical Action's project intervention has increased the fuel choice options of small-scale brickmakers, which may help to ensure their viability in the harsh socio-economic conditions that prevail in Zimbabwe. There is still much to be done to help brickmakers produce better quality bricks that will command a better price, however. The same is true of increasing energy efficiency, which would benefit both livelihoods and the environment.

CHAPTER 7
The view from Europe

Ray Austin

As a glimpse into the potential future development of some small-scale brickworks in the majority world, it may be illuminating to look at the waste application technologies being used and researched in 'the West'. This chapter reviews what is going around the world, focusing particularly on Europe. It concludes by assessing a number of waste application technologies with current or future interest to small-scale brickmakers and those working to support the sector. Over the years the success of a brickmaking unit, regardless of size, has generally been dependent on the following criteria:

1) the ability to provide an acceptable building material appropriate to meeting the needs of the user and the requirements of local standards;
2) close proximity to the market and supplies of raw materials and resources;
3) acceptable standards achieved at the lowest manufacturing cost.

We could summarize these criteria as product, proximity and price; all must be right. For small-scale brickmakers, using appropriate technology and requiring the minimum of capital outlay are usually determinants of producing bricks at a saleable price. Even in areas of Europe where there has been a long tradition of brickmaking and enterprises are typically larger, it has not been easy to achieve a viable brick plant when there are almost constant changes in markets: competitors innovate and undercut, labour is drawn to other industries where wages are higher or conditions better, there are swings in demand for bricks and competition for resources. To achieve a viable unit, achieving product quality at a saleable price has meant being adaptable and opportunistic, no more so than in the search for raw materials, fuels and usable waste materials. The addition of environmental standards in more recent times has made providing an acceptable product at a competitive price an even more challenging task.

An early example of a viable system with a claim to environmental sustainability was the manufacture of bricks along the southern shore of the Thames Estuary. These bricks were used in the construction of London in the 1800s and early 1900s. They were delivered to the capital city in as many as eighty 'sailing barges', vessels that could sail up the Thames and also navigate the canals and rivers flowing into the Thames. On the return journey the barges would load up with refuse from a site in East London and take this back to the brickworks. There it was screened to achieve a granular mix of cinders and combustible material and added to the clay mix as a fuel source. Also added to the clay mix was the waste from the local factory producing 'whiting'. This waste contained chalk and mica and was beneficial in the brickmaking process. Barges would also go out into the estuary and dig sand from the sandbanks at low tide. This sand was used for de-moulding bricks. The brick-earth itself was dug locally in such a way that the fields were reinstated with topsoil: the field would end up lower but still capable of growing crops. Originally, bricks were made by hand and dried in outdoor 'hacks'. As demand grew machinery was installed and driers built. This system of using waste and local resources continued for nearly a century and the low cost of manufacture enabled affordable housing for Londoners and the tide of people swelling its suburbs.

Unfortunately there is not always such a range of low-cost resources available. Moreover, the increasingly high demand for bricks through most of the 20th century has heralded mass production techniques that are designed to burn conventional fuels. Such fuels have been quick and convenient to burn, requiring little or no preparation. For most of the century in Europe they have also been available affordably and without the stricture of regulation applying to greenhouse gas emissions.

Table 7.1 Costs (UK, 2006) of fuels and wastes

(a) Fuel	*Cost per MJ (pence)*	*Relative cost (coal = 1)*
Natural gas	0.570	2.19
Oil	0.720	2.77
Coal (ex pit)	0.260	1.00
Metallurgical coke (ex works)	0.510	1.96
Wood	0.220	0.85
Electricity	2.220	8.54
(b) Waste-as-fuel		
Rice husk	0.077	0.30
Bagasse	0.088	0.34
Town ash (high calorific value)	0.089	0.34
Paper waste	0.107	0.41
Town ash (low calorific value)	0.129	0.50
Coal slurry	0.142	0.55
Sawdust	0.177	0.68
Straw	0.203	0.78
Metallurgical coke	0.517	1.99
Boiler ash	0.417	1.81
Fly ash	2.000	7.69

Every fuel listed in Table 7.1(a) can be used as a direct fuel in either a continuous or batch kiln with the aid of burners or trickle feed hoppers. Coal and coke can also be used as additives in the clay mix. In most kilns, thus incorporated fuels can supplement the direct-fired fuel. In the case of a clamp, Scotch Kiln or Vertical Shaft Kiln (VSK), coal and coke in the clay mix can provide the entire energy requirement. With the exception of the VSK, which has a high efficiency, the body-added fuel is only 35 to 60 per cent as effective as the direct-fired fuel in usefully using its heat. Effectiveness cannot be directly equated with thermal efficiency, however. Thermal efficiency can be boosted by the incorporation of wastes in the clay mix.

Table 7.1(b) does not include the cost of transporting the waste product from its source to the brickworks. This can add considerably to the cost per MJ when estimating the feasibility of its use. As we have previously noted elsewhere, with some waste materials there are benefits other than being a source of energy. They can also reduce wastage when drying the bricks by reducing shrinkage and by opening up the clay to enable the moisture in the brick to migrate to the surface. This reduces differential shrinkage within the brick and reduces the risk of cracking. Really sticky, highly plastic clays in particular are prone to drying cracks and up to 10 per cent of the dried bricks can be unsuitable to progress to the firing stage. Boiler ash, fly ash, sawdust and shredded paper in particular can improve the drying process.

European brickmakers have a long history of adapting to available fuels and have moved from wood to coal, to heavy oil and most recently gas. Sometimes these changes have been of necessity, as in the instance of wood scarcity. Otherwise changes in fuel choice have been due to environmental concern, i.e. the change from oil or coal to natural gas. Natural gas prices in Britain have recently doubled, however, and the quest for alternative fuels is once again underway. Bearing in mind that climate change is an increasingly significant factor, the search has focused on waste gaseous fuels because these burn with less carbon dioxide emissions. Already, 'coal-mine methane' has been used successfully. This gas can be extracted from either working or abandoned mines. If the methane were vented directly to the atmosphere as a waste, it would have

Box 7.1 Incorporation of wastes

Body-added fuels can be used more efficiently by optimizing the way bricks are loaded into the kiln, ensuring there is accessibility of oxygen to the fuel. Recent trials carried out in clamps have reduced the firing time by 22% and wastage has decreased from 8.7 to 7.8%. This has been achieved with a reduction of body fuel energy from 4,346 MJ per 1,000 bricks to 4,028 MJ per 1,000 bricks. The bricks have been set in a vertical straight pattern, allowing oxygen to reach the bricks from the top to the bottom. Any loose sand was removed from the bricks before setting in the clamp to prevent sand affecting the oxygen accessibility. The fuel bed on which the clamp is set burns cleaner and quicker if it is a fuel with a low ash content.

twenty times the effect as the same amount of carbon dioxide on the greenhouse effect. Another gas source exploited by brickmakers has been landfill gas. This is produced by the decomposition of domestic, industrial and agricultural waste dumped in the worked out areas of the clay pit adjoining the brickworks. In both instances location – proximity – is a critical factor with respect to viability. It is not generally viable, for example, to extract, store and transport coal-mine methane by road. Meanwhile, electricity generating enterprises also find these gases an attractive alternative and so are in competition with brickmakers.

Another change affecting the brickmaker is the trend towards clay bricks and blocks with properties that reduce heat loss in buildings and minimize the use of mortar. This trend has focused on the manufacturing of lightweight units, which are structurally sound and have excellent thermal insulation characteristics. German brickmakers in particular have undertaken extensive research into using sawdust and also shredded polystyrene foam derived from used packaging. Such materials are added to the clay and bricks are extruded using a die. After firing, the result is a light-weight brick or block that has many small pores where the sawdust and polystyrene foam has burnt away. The sawdust also contributes energy to the firing process, of course. The use of clay is minimized, moreover, thereby reducing the energy requirement to fire each square metre of walling material. Paper sludge is also used to similar effect. Unfortunately, a conventional direct-fired kiln is required to fire these products at present, rendering them of little current interest to small-scale brickmakers in the majority world.

Mining waste has often found a use in brickmaking, especially heaps of coal-mine waste such as the 'bings' of southern Scotland and the waste heaps of the Midlands, the Ruhr and Poland. These shales have a high content of carbonaceous material and, in some instances, this may even be sufficient to fire bricks without further fuel addition. Meanwhile, South African gold mining gives rise to waste known as mine slimes which, when added to clays, enhance the fired product at virtually no cost.

It cannot be over-stressed that the cost of transporting waste any distance can make its benefit marginal. Frequently, though, the person creating the waste material has problems with storing and disposing of it, and disposal can often be equally expensive. In some circumstances, such as with fly ash, boiler waste, mine waste, paper waste and sawdust in some regions of the UK, the waste producer may deliver the waste to a local user free of charge. The partnership of waste producer and waste user can prove extremely beneficial to both parties.

As circumstances change and a waste may become more valuable for generating electricity or fuelling boilers, a pragmatic approach is needed. Competition is almost inevitable as all industries look to cut costs. Apart from being used as fuels and additives in brickmaking, for example, sawdust and wood waste can also be converted into board. Fly ash can be used to make lightweight concrete blocks, moreover. The search for waste products to improve the brickmaking process by introducing pore-forming materials has followed numerous trails. Very often, however, there is an alternative market prepared to pay more for the waste as, say, a boiler fuel or an animal feed additive. Vigilant brickmakers are constantly on the lookout for 'spoil heaps' where useful materials have been (poorly) incinerated, buried or dumped. Those civil projects that require large excavations, for instance, can be disposing of usable clay at no benefit to anyone. Some years ago, thanks to the keen eye of the owner of a brickworks who noticed it was being dumped, the clay excavated from a London underground railway project was diverted to a brickworks at no cost.

If they are situated adjacent to a brickworks, the waste heat generated by other industries can be diverted and used in brick driers. Naturally, this is especially welcome in the wet season. A brickworks near Bristol was located next to a factory producing 'carbon black', which is used to as a reinforce products such as tyres, tubes, conveyer belts, cables and other rubber goods. The factory flared off 'Jones gas', a by-product of the production process that contained carbon dioxide, carbon monoxide, hydrogen and water. This gas was readily piped to the brickworks and fuelled the kiln there for many years.

Environmental concerns

European brickmakers are very conscious of four areas of environmental concern:

1. greenhouse gas emissions, notably (i) carbon dioxide which is generated from the combustion of carbon fuels such as coal and (ii) methane generated from bacterial decay of organic matter (agricultural waste, forestry waste, landfill and coal mines);
2. dioxin and furan (a group of colourless, volatile, heterocyclic organic compounds containing a ring of carbon atoms and one oxygen atom) emissions from the combustion of certain plastic waste, such as PVC and other chlorinated plastics, which can pose a risk to human health;
3. heavy-metals and pathogens found in sewage sludge that can cause them to be unsuitable for spreading on agricultural land and can be a health hazard if handled without due care;
4. leachates from landfill waste sites that can seriously pollute watercourses and aquifers.

There are instances when using waste can reduce negative environmental impacts. The use of methane as a fuel, for example, is preferable to allowing it to escape into the atmosphere as methane. The carbon dioxide produced in methane combustion contributes much less to the greenhouse effect. Though more research is required, there is the potential for toxic heavy metals to be used as additives in brickmaking. As we noted in Chapter 3, it may be that these heavy metals can be chemically locked into the brick during firing, thus rendering them harmless.

There are other areas of concern such as smell, which may be harmless to health but still be considered unacceptable. Vermin, associated with landfill, paper waste and agricultural waste in particular, can be a significant, though usually local, problem. Water run-off from sites that store waste in the open can contaminate watercourses and wherever possible should be stored on higher ground and covered.

Waste materials and brickmaking technologies

Biogas is a mixture of methane and carbon dioxide. It is produced by bacterial digestion of organic matter and can be used as a fuel. Anaerobic digestion involves placing biodegradable organic materials, the feedstock, in an environment absent in oxygen. Depending particularly on the nature of the feedstock, the process can give variability in the gas produced. Wastes are usually digested anaerobically in a closed fabricated container such as a drum. Gas production is temperature-dependent and requires a minimum of 10°C. Up to a limit, increasing temperature accelerates natural digestion and gas is produced and stored, usually in a chamber designed to maintain a constant pressure. Despite being associated with small-scale domestic gas supply in China and also dairy farming in Europe and USA, where the gas is often used in electricity generation, this technology appears underdeveloped in many regards. Research in India has investigated the use of anaerobic digesters in generating gas to assist the drying of clay tiles.

As feedstock for anaerobic digestion, dung, sewerage and many agricultural wastes are good. Sawdust and straw are not so good. Typically digesters need daily attention and conditions inside should be kept reasonably stable. Due to potential explosion risk and lack of breathable oxygen, care is necessary when carrying out any cleaning or maintenance to digester vessels.

Producer gas consists of a mix of mainly nitrogen and carbon monoxide. It has a low calorific value because the large nitrogen constituent is inert. Producer gas is made in a furnace or generator. Air is forced upward through a burning fuel. Although the fuel is introduced from the top, no air is admitted there.

Box 7.2 Biogas, producer gas and landfill gas

Biogas

Analysis: 57–70% CH_4, 30–40% CO_2, 1–10% N_2, 0–1% H_2, traces of O_2 and H_2S.
Calorific value of raw biogas: 23 MJ/m^3.

Producer gas

Analysis: 40–50% N_2, 22–27% CO, 10–15% H_2, 10–15% CO_2, 2–3% CH_4.
Calorific value: 4–6 MJ/m^3.

Landfill gas

Analysis: 57% CH_4, 42% CO_2, 0.5% N_2, 0.2% H_2, 0.2% O_2, traces of H_2S.
Calorific value: around 19 MJ/m^3 at its peak.

Carbon in the fuel is oxidized by the oxygen in the air to form carbon monoxide. The nitrogen in the air is unchanged. When steam is introduced with the air, the producer gas will also contain hydrogen. Producer gas is quite widely used in industry because it can be made from inexpensive fuels and is more versatile. The only well-documented example of the use of producer gas in brickmaking, it seems, comes from Finland. There, a tunnel kiln used extruded peat (densified) in step grate converters to produce gas to fire clay blocks. Such a process requires considerable investment and technology development, however.

Waste biomass is quite widely used on all continents, especially when conventional fuel is in short supply or expensive. The calorific value typically ranges between 14 and 19 MJ/kg, comparable to fuelwood. Handling and storage are safe and reasonably straightforward. It being a potentially renewable source of energy, burning biomass is relatively environmentally friendly. Generally, the ash content and emissions of sulphur are low, and burning reduces the natural methane that evolves from rotting crop waste. When in a granular or powder form, such as sawdust, coffee husks, shredded paper or crushed shell, biomass can be used as an additive to the clay mix.

Alternatively, it can be formed into fuel briquettes, either by hand, by pressing or extruding. Typical sources of biomass for briquetting are crop waste or forest debris. The density of briquettes varies depending on the method of making. Generally denser briquettes are preferred. With low-pressure briquetting, either by hand or pressing, it is often difficult to attain a product that holds together, particularly when using springy biomass. In such cases a binder is necessary and an appropriate substance is not always easy to source cost-effectively. Some binders, such as clay, leave a substantial residue to deal with; others, such as cement, are relatively expensive for such a use.

Fluxes are used in a fine dust grading and added to the clay mix to reduce the maturing temperature of the brick. They are usually a waste by product of a process such as mining or quarrying. Recently, ground waste glass has been used with some success. It is rarely economically viable to grind fluxes for use in brickmaking, however, as the mechanical energy used exceeds the thermal energy saved. Fortunately, many mines and quarries create dust, which they are obliged to dispose of. In particular, road-stone material, such as granite and basalt, when added to the clay mix as dust mean that a lower firing temperature is needed to achieve a satisfactory brick. These additives can, in some instances, also reduce the drying shrinkage and waste from drying cracks. Wood ash is one flux that offers this advantage.

Using a flux addition can affect the vitrification range. If this range is small, there is a risk of the bricks being either under-fired or over-fired. The top temperature in a clamp could be anything from 950°C to 1,050°C and if the vitrification range of the clay/flux mix is 1,000°C plus or minus only 20°C, there could be a poor yield of good bricks. Fluxes are especially useful when the only clays available are refractory and have a very high firing temperature. Furthermore, if the fired bricks are to be used in a very cold climate and are therefore susceptible to frost damage, then the use of fluxes is beneficial.

One tonne of degradable rubbish can produce 400 to 500 cubic metres of landfill gas, although this is only some 20 per cent of the potential energy that could have been recovered if the waste had been incinerated. Due to clays tending to be being impervious, when 'worked out' brickwork quarries have become a favoured location to dispose of municipal solid waste. Very soon after this means of disposal was employed, however, it was noticed that methane was being evolved and harming trees. Future landfill sites were arranged in such a way that this methane could be flared off and so rendered harmless, in respect to local flora at least. As this was frequently occurring adjacent to brickworks, the practical use of this methane was investigated. Following trials, it was used in the hot zones of the brick kilns. An alternative is to burn methane to generate electricity for use either on or, indeed, off the site.

Some sites in the UK have now been using landfill gas from the original dump site for 20 years. The volume and calorific value of the gas varies over the years. The volume reaches a peak and then declines to zero when all the organic material has been digested anaerobically. The calorific value also reaches a peak and then tends to stabilize until the site stops producing. Present-day landfill sites are designed with great care, being laid with a network of pipes to collect as much gas and as little air as possible. The pits are lined and the waste capped with plastic sheet after water has been added to aid decomposition. This results in a leachate, which must be collected and treated to neutralize its potentially contaminating effect on aquifers and streams. Landfill gas is corrosive and any metal pipe-work, valves and regulators require regular examination, maintenance and replacement. Extraction of landfill gas also requires constant monitoring and attention by an operator.

Shredded paper is commonly used as an additive to assist brick drying by 'wicking' moisture to the brick surface. Moreover, paper additive bulks out the clay and provides a source of energy during firing. As detailed previously, the addition of paper can also improve the thermal insulation properties of lightweight clay blocks and this technology is employed in Germany, Switzerland and Austria, in particular.

Also used as an additive is 'incinerated paper ash', which is the residue after waste paper has been burnt in a boiler or incinerator. Though it has virtually no residual energy, paper ash does assist drying and, even in small additions such as 2 per cent by mass, can reduce the tendency of the brick to crack during drying. Many types of incinerated paper contain calcium carbonate, kaolin and titanium oxide, which can affect the fired colour of the brick by bleaching red to a lighter colour and also giving white spots. Although harmless, customers may perceive such 'blemishes' as indicative of poor quality.

Paper recycling is now commonplace and paper fibres can be reused several times before they become brittle and unsuitable. The fibres are extracted from the recycling process and the residue is known as paper sludge. This sludge can be further refined into dry fibre or dry filler (calcium carbonate, kaolin, titanium oxide and ink residues) and even compacted into dry fibre briquettes. The fibrous material has a gross calorific value approaching that of wood and when added

Box 7.3 Dioxins

Dioxins are a family of toxic chlorinated organic compounds that bioaccumulate in humans and wildlife over time because they are soluble in lipids (fat). The most notorious of those is 2,3,7,8-tetrachlorodibenzo-p-dioxin, often abbreviated as TCDD. Even at low exposures, dioxins can accumualte to dangerous and even lethal levels.

Source: Wikipedia, http://en.wikipedia.org/wiki

to clay reduces the risk of cracking during drying. The combustion of the fibre in the clay body happens at relatively low temperatures and, unless the clay has a very low vitrification temperature, it needs to be supplemented with other fuels, typically coal-dust or screened ashes. Extruded briquettes are at present only available in North America and Europe due to the cost of the high-pressure equipment required.

Refuse Derived Fuel (RDF) has a high paper and cardboard content and is derived from domestic waste. RDF is converted to briquettes and pellets, usually by extrusion, and is then a convenient form of solid fuel. Caution is required if there is any plastic content as this may give off harmful fumes when burnt. The caloric value varies from 12 to 17 MJ/kg, depending on the source waste and the nature of processing.

Plastic Fuel Pellets are formed in municipal waste recycling units where the disposed items are split into paper, glass, metal and plastic. The plastic fraction is shredded and extruded into pellets, which can be used as fuel. These have been tried in the high-temperature zones of brick kilns with a limited degree of success. The pellets are fed through the roof of the kiln and combust at temperatures in excess of 1,000°C in the firing dykes. Due to the handling involved and the lack of control, however, this method is not entirely satisfactory. There is concern that dioxins and furans are formed at lower temperatures (300/400°C) and, although 'destroyed' at 1,000°C, they can reform if cooled quickly. Both dioxins and furans are persistent and potentially harmful to humans. Plastic should not be used as a fuel unless constant monitoring and control is exercised in the feedstock used, the process of combustion and emissions.

Sludge from waste-water treatment plants is normally treated with lime, dewatered and disposed of on land. Sewage sludges may contain heavy metals and so disposal on land is controlled. They may also contain pathogens, which can be harmful if handled without adequate protective equipment. In dense population areas, particularly, sewage sludge poses a significant disposal problem. Hence, there is some research into using dried sludge or sludge ash in building products. As an additive to fired bricks, up to 40 per cent of dried sludge and 50 per cent of sludge ash are the limits, with ash yielding the higher-strength product. Benefits are an increase in clay plasticity and, in some instances, reduction in drying shrinkage. On firing there is a tendency for shrinkage to increase, however. (Fired bricks typically shrink twice, at the drying and the firing stages, the former almost always being most significant.) The process of firing may also produce an unpleasant smell. The fired brick will have a marked increase in water absorption and lower compressive strength, as well as a tendency to effloresce. When mixed with clay, sludges with a high lime content will have a short firing range and it is not advisable to use a firing process that has a high temperature variation. The caloric value of raw sewage sludge is around 23 MJ/kg.

Burning used tyres directly is neither permissible, advisable nor convenient, particularly not in contemporary brickmaking in Europe. Burning tyres cause air pollution, endangering human health, flora and fauna in the vicinity. Hence, such a process is prohibited by the environmental legislation pertaining in EU countries. Moreover, tyres are not a form that would be convenient for use as a fuel at most brickworks. There are currently at least two companies investigating the feasibility of converting old tyres into a high-carbon granulated fuel, a type of char. Over the last three years, they have come close to achieving a feasible operation. The product is still not on the market, however, and the chemistry of the process is still under development.

Wastes with potential for small-scale brickmaking

It is suggested that, for burning waste to be a beneficial technology for small-scale brickmakers, a number of criteria must be met:

1) The cost of the waste in a usable form to the point of use is economically viable.
2) The capital cost of any equipment required is low.
3) The technology is simple, repeatable and safe.
4) The supply of the waste material is guaranteed over a reasonably long term.
5) The quality of bricks produced is good and there is a high yield of saleable bricks.
6) The net environmental impact is negligible, beneficial or can be mitigated.

Granulated agricultural and forestry waste added to the clay mix meets most, if not all of these criteria. This is reflected by the fact that such fuels have indeed been adopted in many brickworks worldwide. For many years, rice husk has been used with considerable success in developing countries as a supplement with wood for direct firing and as a body additive in clamp firing. It has also been used to manufacture a lightweight refractory insulating brick for use in kiln and furnace construction.

Fuel briquettes and pellets derived from agricultural waste, forestry waste or waste paper may also meet our criteria and are very promising technologies. In some part of the world such fuels have found favour in the direct firing of kilns or in the fuel bed of a clamp. Sawdust, rice husk and cotton stalks are difficult wastes to use as energy sources. Whilst they can be fed into a hot kiln fire-hole, they have to be used sparingly to prevent smothering the fire. There are burners that will blow air and pulverized fuel such as sawdust into a kiln when it has reached temperatures in excess of 800°C. Below this temperature these burners are ineffective, however, and they also require an electricity supply. Compacting these agro-forestry wastes into briquettes means these problems can be overcome and they can be used for direct firing kilns, either independently or supplementing other fuels. With clamp firing, they can be used on the clamp bed. The calorific value is similar to that of wood and, depending on the waste used, burning can give very little ash.

There are numerous briquetting machines available, many developed in India and China. Apart from manual presses, extrusion by a reciprocating ram or an Archimedes screw are the usual alternatives. The use of binders is not always necessary when high pressure and a heated die are used to extrude the agricultural residue, especially if sawdust is added. The alternative to high pressure is adding a binder such as clay, though this usually results in inferior combustion. In spite of biomass briquettes being used successfully in many regions, there is a need for 'best practice' to be shared to perhaps simplify and improve the reliability of the briquetting equipment. High wear on the screw and the die are not easily resolved in a low-technology environment. With high energy prices hitting Europe and East Asia, along with international commitments to reducing greenhouse gas emissions, biomass briquettes and pellets are seen as serious alternatives to coal and oil. There is a strong case for examining existing technology and highlighting opportunities to make these briquettes feasible in both low- and high-technology environments. What, for instance, are the best shape and size and density of the briquettes for different applications? And which conditions give the most efficient combustion?

In many majority world countries, not only could this fuel briquetting technology benefit brickmakers, it could also be employed to produce suitable fuel for domestic cooking. Initially, the briquettes would be simply made by hand, possibly using paper soaked in water as a binder. In the future, as appropriate technologies are developed, briquette quality and economic viability could be improved and the market expanded. There is a strong case for making bricks and fuel briquettes on the same site, as both processes require similar skills and the latter could generate additional employment and community income.

Boiler ash and mature domestic waste, which from a health and safety point of view should be more than thirty years old, can also meet most of the criteria. Shredded paper or paper sludge can largely be regarded as renewable sources of energy. The cost of processing along with the care required in handling and use make such wastes mainly unsuitable for consideration by small-scale brickmakers. There is some scope for developing anaerobic digesters and landfill gas extraction for use in small-scale brickmaking in the not too distant future. Indeed, these are technologies that deserve planning and initial implementation as soon as possible.

CHAPTER 8

The feasibility, sustainability and viability of using wastes

Kelvin Mason

> Sustainability is an ultimate, perfect condition, which we will almost certainly never attain: a perfect condition in which the human species manages to live within the (planetary) limits that we've got in a way that enhances and protects the diversity of what we've got... Sustainability's also a world in which societies are stable, where there's a high level of social equity, where societies are peaceful... And it's also a place where there's an economy for all; there is enough for all people. So, it's a highly idealistic notion of the way that we could live. But as a vision, it's worth holding up. (Rod Aspinwall, UK Sustainable Development Commissioner, quoted in Mason, 2005)

In this final chapter we will assess the technical feasibility of using wastes in brickmaking. We will also ponder whether doing so is more or less environmentally sustainable than alternatives. Some of this technical and environmental ground will have been provisionally covered in our country case studies. Herein, then, we will attempt to develop what we have already noted. We will also consider the institutional conditions under which the use of wastes is, or could be, economically viable and so boost the livelihoods of brickmakers. Such conditions will be influenced by the political support that might be won for the technologies under consideration. Finally, taking all these issues together, we will speculate upon the actions that policymakers, NGOs, fieldworkers and brickmakers could take in the light of our deliberations.

Technical feasibility: we can do it

In the main, we have considered using agricultural residues and industrial wastes as fuels in brickmaking. Whilst we have touched upon the use of such materials as fluxes, grogs and bulkerizers, and also upon using brickmaking as a means of waste disposal, our conclusions must largely be confined to fuel substitution. It is apparent, both from the literature we have reviewed and also Practical Action case studies from around the world, that the use of wastes and residues as fuel substitutes in brickmaking is, generally speaking, technically viable. Though one conclusion of our investigation must be that each instance of potential substitution should be assessed on its own merits in its own particular context, it is safe to confirm the enormous potential energy available from wastes and residues.

Although some wastes can be burned in fires beneath clamps or kilns in a similar manner to fuelwood, other means of combustion offer significant advantages in many cases. Incorporating finer residues in bricks, for example, is a way of putting the fuel in intimate contact with the clay. It can, therefore, be a more efficient means of combustion than burning the residue remote from the bricks, minimizing the heat loss associated with transfer over distance. Another advantage is that the bricks produced will be lighter because the fuel fraction of the waste will have been burned away. Lighter bricks with more voids are cheaper to transport, easier to handle and offer improved thermal insulation.

The limit on how much fuel can be incorporated into bricks is set by the effect on both the handling strength of the green brick and also the required properties of the final product. When considering incorporating wastes, the mouldability of the clay, the green strength of the brick and the durability of the final product are particular concerns. The addition of most wastes reduces the plasticity of the clay mix and thence the green strength and ultimately the durability of the brick. In general, though, the finer a residue, the greater the amount of it that can be incorporated. Results from the Sudan case study suggest that, as a rough practical guide for fieldworkers, the final mix must contain at least 30 per cent clay by mass. Other studies have indicated that the production of sufficiently durable bricks means the fuel fraction of the mix will be between 5 and 10 per cent by mass. It should not be forgotten that the appearance of the final brick, its acceptability to customers, may be more important than its physical properties when brickmakers' livelihoods are at issue. I suspect many a frustrated engineer has, like me, solved technical problems only to confronted by the fickleness or conservatism of consumers: people just don't know what's good for them!

In some cases too, it is the producers who baulk at technologically sound innovation. In Zimbabwe in the early 1990s, for example, the design, testing and production of a sand-moulding table offered brickmakers the opportunity to produce bricks that reached 'common' quality. By virtue – or otherwise – of the restrictive standards in force, such bricks could be sold to the more lucrative urban market of the time. The moulding table was made from local materials using local skills and was therefore available and affordable. The technology of using the table was certainly not too alien for brickmakers to adapt to. Although the bricks produced could compare favourably with the output of the IT Workshops moulding machine, for example, the perception of brickmakers was that the technology as a whole was inferior to such imported alternatives. So, they did not adopt the moulding table readily and en masse, as might have been the case had they judged the matter wholly rationally, i.e. in the way that engineers and scientists like to think these things are decided.

The psychological perception that imported products must be better than locally made ones and that mechanization is always superior to processes that are labour-intensive is widespread and deeply held. In the majority world, images of the benefits of the Western model of industrial modernization are pervasive and persuasive: fast cars, cheap air travel, consumerism... The jury is out on whether any of us will successfully make the transformation to ecological modernization. What that model will look like is still being contested, in fact, though many of us believe the time for argument about formative initiatives such as the Kyoto Protocol is long past.

Returning to our central theme of technological effectiveness, placing waste-as-fuel in layers between rows or columns or bricks in clamps or kilns is another means of bringing the fuel closer to the brick. In this case, however, there is no unavoidable effect on clay or brick properties. One limit on using waste-as-fuel in this way is likely to be the danger of insufficient air-flow and hence incomplete combustion. Finer wastes, in particular, may well compact to the extent that airflow is restricted. This may lead to burning in a reduced oxygen environment and thence to reduction spots on the bricks, possibly lowering their sale value. Incomplete combustion also tends to increase air pollution. Another limit, particularly with respect to clamps, is the structural integrity of the stacked green bricks. Generally, brickmakers are known to build clamps that sag inwards as any fuel contained within them burns away. If, as the waste-as-fuel burns away, any resultant movement in the stack of bricks is not controlled, then there may be a risk of total or partial collapse and hence loss of production and possibly accidents involving staff. With certain wastes, notably the by-products of rice production, a high ash content may present technical problems with both combustion and ash disposal.

One promising means of preparing certain wastes for optimal burning falls under the label of low-pressure briquetting. In Peru, coal-dust and clay were formed into spherical 'briquettes' by hand. The amount of clay used is just sufficient to bind the coal-dust so that the briquettes hold their shape and can be handled when dry. The construction of rudimentary grates made from bricks makes it possible to burn these briquettes in tunnels below the kiln. This means the technology is not significantly altered from burning fuelwood and no capital investment is required. Burning briquettes in this manner is combined with both using coal-dust between layers of bricks and also distributing briquettes at known 'cold spots' to promote even firing of bricks throughout the kiln. It is worth noting here that, in another Practical Action collaborative project, brickmakers in Ecuador developed the technique of placing fuelwood in their kilns at similar peripheral locations where bricks tended to be under-fired.

In Sudan, meanwhile, rotted bagasse was made into blocks with the aid of a manually operated press. Technically, this operation too falls into the category of low-pressure briquetting. A range of binders was identified and tested. Unfortunately, molasses, the most promising binder technically, was not available in the quantities and at the price that would have made its use in block production commercially viable on a mass scale. It was acknowledged that more research on clay and filter cake was required. As filter cake is available as a residue from sugar factories, it would seem potentially ideal as a binder for bagasse produced at these locations. On the other hand, if bagasse were available in its loose form in close proximity to a brickmaking site, then there would evidently be a ready supply of clay to use as a binder for fuel briquettes. Given the appropriate soil properties and processing technology, a machine such as the Ceratec press tested by Practical Action in Zimbabwe could be employed to press both bricks and also the fuel briquettes to fire them. Press machines of a similar design are available from other manufactures. The main drawback for small-scale brickmakers in much of the majority world, though, would consistently be the purchase price of an imported piece of hardware.

Overall, the briquetting of wastes offers a number of benefits. Loose and low density wastes-as-fuels can be rendered easier to handle and to transport. Moreover, briquettes can be burned more readily, and generally with less technological change, than can powdery residues and wastes. They also offer a way of overcoming the limit on how much powdery waste can be burned by incorporating it into bricks. The combination of burning a powdery waste incorporated into clay, formed into briquettes and also distributed in the kiln may well be a technological route to the complete substitution of traditional fuels such as wood or coal. It is also a means of securing substantial gains in energy efficiency.

Environmental sustainability: we should do it

From the outset we proposed a quite complex definition of environment that is doubly inclusive, i.e. that both includes and also is included in humanity. Part of what makes us human is nature, and nature is in part humanity. Philosophically, this leads us away from an econometric view of the environment. Nature may provide us with, in the economist's terms, goods and services, but the stewardship ethic means we have a duty of care to ensure that nature is not destroyed or degraded. This ethic holds whether or not the element of nature considered renders goods and services that are valuable to, or rather valued by, humanity. There is a difference. For example, the greenhouse effect is extremely valuable to humanity but it becomes a problem, some critics would say, because it is not valued by us. Measures such as carbon taxes literally put a value on the greenhouse effect. The stewardship ethic therefore encompasses the principle of intrinsic value. Though it can be problematic to put into practice, intrinsic value need not mean that all change is taboo. Rather, it is a route to considering the environment beyond that which can be measured quantitatively in units that reflect only short-term, transient anthropocentric value.

Sustainable development is essentially a matter of space and time. Conscious of the needs of brickmakers and their families in today's world, we must also try to take account of the needs of their descendants and, indeed, our own. It is, as the popular contemporary phrase runs, a very big task. Through Finn Arler's typology of resources and adapting the methodology of Environmental Impact Assesement (EIA), we came up with a means of considering changes in brickmaking technology. Our grandchildren or great-grandchildren may judge us on the stock of exchangeable, critical and unique resources that they inherit.

Box 8.1 Goliath

They chop down 100 foot trees
To make chairs
I bought one
I am six foot one inch.
When I sit in the chair
I am four foot two.
Did they really chop down a 100 foot tree
To make me look shorter?

(Milligan, 1973)

Even when we consider only the economic microcosm of small-scale brickmaking, all these resources are potentially affected. Brickmaking may be using up exchangeable resources in the form of fossil fuels, degrading unique landscapes, or contributing to the destruction of the critical ecological service rendered by the greenhouse effect.

When we considered the sustainability of small-scale brickmaking in Chapter 2 we prioritized a number of changes in our strategic EIA, namely deforestation, emissions of carbon dioxide, emissions that pollute air and affect air quality, the incineration of wastes, and flora and fauna. As long as brickmaking is not degrading a unique landscape, I suggest that this list holds good in the light of what we have gleaned from the intervening chapters. Applying our adopted typology of resources, we can see that deforestation, emissions of carbon dioxide and air pollution pertain to critical resources, as might the incineration of waste. Flora and fauna may be considered either unique or exchangeable resources, meanwhile, depending on their individual characteristic.

As a benchmark, I propose that, from an environmental point of view, the best possible source of fuel for brickmaking would be biomass from a sustainably managed local source. If this biomass were the by-product of an agricultural process that produced a primary crop for consumption or use by humanity, so much the better because I believe this would make our duty of care to dispose of that waste in an environmentally friendly fashion even clearer. My proposal, however, ignores the fact that wastes and residues have already been produced as a result of unsustainable production, e.g. boiler ash from an inefficient power station in Harare and sawdust from commercial operations in Peru. Until the nature of such production has been amended, it may well be that one of the best ways of disposing of wastes and residues is via brickmaking. Before we get too general, though, let us consider three examples from our case studies and assess the environmental impact of their dissemination on the national scale.

The case of rice husks in Peru can, I propose, be considered as fairly representative of agricultural residues. As we have discussed, the use of each residue in each context must be assessed on its individual impacts. Nevertheless, considering the environmental impact of using rice husks according to the priorities we have established should at least enhance our wider understanding.

On the national scale, the use of rice husks as a substitute for at least a proportion of fuelwood would have a positive environmental impact on deforestation. Trees are a critical resource, serving as carbon sinks that can help ensure the greenhouse effect is life-giving rather than life-threatening. Burning rice husks is also carbon neutral in that emissions of carbon dioxide can be no more than that which the plant absorbed whilst growing. Moreover, rice is an annual crop and so is, in that sense at least, sustainably produced. If rice husk is incorporated into bricks and burned in layers in the kiln, the process will tend to be more efficient thus further reducing carbon emissions. All in all, then, the dissemination of using rice husks as a fuel in brickmaking on the national scale would have a positive impact on the global and regional environment.

With respect to emissions into air, the overall quantity of emissions is unlikely to change significantly. Rice husks have a similar calorific value to fuelwood and so a similar amount will be burned to produce the energy required. In terms of air quality, however, it may be that burning rice husks emits proportionally more acid gases and particulates in smoke than wood. Generally, this effect on the local and regional environment would need to be assessed on a case-by-case basis, comparing a particular wood with a particular agricultural residue. Certainly, the nature of rice husk will make it more technically difficult to burn cleanly, i.e. to approach complete combustion. If rice husks were used as a substitute for coal rather than fuelwood, the environmental impact on both air pollution and net carbon dioxide emissions would be likely to be positive, depending on how cleanly and efficiently the fuels were burned.

Vis-à-vis disposing of waste, rice husk is a residue with a limited number of alternative uses. Although it does not seem to present a major disposal problem in most instances, rice husk can be incinerated in brickmaking. Moreover, the energy released is used productively. If rice husk is used extensively in brickmaking, however, there may be an increase in heavy vehicle traffic for its transport. This is likely to have a negative impact on roadside flora and fauna, though the magnitude of change will probably be small. Assessing the impact on the local environment would mean balancing the benefits of productive waste disposal against any damage caused by increased traffic. Mitigating the extent of such damage may be possible via reasonably straightforward measures. Were it the case that rice husk was transported extensively by road, the increase in emissions of carbon dioxide and air pollutants from delivery vehicles would have to be considered.

Let us now consider the environmental impact of disseminating the use of boiler ash in brickmaking on the national scale in Zimbabwe. To an extent, the following discussion will give us insights into the use of similar industrial wastes – other ashes and coal-dust – in alternative national contexts. As with all alternative fuels considered, provided that it is fuelwood that would otherwise be burned, using boiler ash in brickmaking has a positive environmental impact, reducing deforestation. Burning boiler ash is not carbon-neutral on the same time-scale as burning biomass, however, whether that biomass is fuelwood or agricultural residues. The benefit of leaving trees standing as carbon sinks has to be balanced against the carbon emissions from burning boiler ash. In the short to medium term, boiler ash would not degrade and emit carbon dioxide if left to stand. On the other hand, using boiler ash in layers between bricks is likely to be more efficient than burning fuelwood beneath a clamp or kiln, and increased efficiency would reduce carbon emissions from the process. Overall, although it is better to burn wood from a sustainable supply, I suggest that burning boiler ash has a less negative environmental impact than burning wood from other sources.

Considering emissions into air, boiler ash and similar fuels will tend to produce more acid gases and particulates than fuelwood. Oxides of sulphur and nitrogen are of particular concern, contributing to not only local environmental pollution but also to acid rain on a regional scale. There may be ways of mitigating these impacts. Developments in clean coal technology could be employed to wash impurities out of boiler ash before burning and also perhaps clean – or 'scrub' - exhaust emissions (BBC, 2005). In general, however, such technologies are currently beyond the reach of the small-scale brickmaking sector.

In Zimbabwe at least, boiler ash is a local landscape pollutant. Initially, it was a waste disposal problem for Harare power station, whose site was simply being overwhelmed with piles of ash. The use of boiler ash as a fuel in brickmaking served to help solve these problems. With respect to burning boiler ash in brickmaking on a national scale in Zimbabwe, the direct impact of road transport on flora and fauna would probably be small. If the ash were transported over long distances by trucks that billow black smoke, as is typically the case in Zimbabwe, then the increase in emissions of carbon dioxide and air pollutants would be a significant impact. There is a rail network in Zimbabwe that reaches a number of population centres, including previously designated 'growth points'. The use of this network to transport a proportion of boiler ash would reduce the impacts due to road transport. Overall, burning boiler ash seems to have the potential to have a less negative impact than continuing to use fuelwood.

Our analysis of the use of bagasse in Sudan actually follows most of the arguments that applied to case of rice husks in Peru. Like rice husks, bagasse is as an agricultural residue. Burning it is carbon-neutral in the immediate term and sugar cane is an annual crop. The use of bagasse as a substitute for some proportion of fuelwood would serve to reduce both deforestation and carbon emissions. In this regard, the regional as well as global environmental impact is positive. There is likely to be little change in air pollution compared to burning fuelwood. If bagasse replaces coal, however, the net impact is most likely to be positive. There is an acknowledged problem with the disposal of bagasse as a waste in Sudan. Thus, the environmental impact of burning bagasse in brickmaking is positive in this respect also. Overall, the case study of bagasse is even more positive than that of rice husks. It seems that the only significant negative impacts might be as a result of transport. While the impact on flora and fauna is likely to be small, the extra carbon dioxide and air pollution associated with distributing bagasse nationally by road could be significant.

Frankly, although it helps with perspective, we do not really need our complex definition of environment or the stewardship ethic to inform our verdict on the case of using of wastes as fuels in brickmaking. It is substantially a win–win scenario. Not only does the use of wastes largely serve to preserve nature, it also enhances the environmental goods and services provided to humanity: forests and associated ecosystems are conserved while there appears to be a positive impact on the greenhouse effect, for instance. A drawback with burning industrial wastes as substitutes for biomass may be a negative impact on air quality. Given the scale and seriousness of the greenhouse problem and considering the other advantages of burning waste, however, the strategic verdict must be positive. The caveat to this verdict concerns the extent and nature of the transport required to provide brickmakers with waste-as-fuel. A second dimension of the win–win scenario is that fuel substitution benefits not only future generations, whose resources are largely conserved, it could also have a positive impact on the livelihoods of contemporary brickmakers. The moral imperative to act to further sustainable development is confirmed. As I have continually cautioned, though, each case of potential waste-for-fuel substitution should be assessed specifically as well as generally.

Economic viability: so what's stopping us?

Before proceeding, I should note that, so far, we have considered two types of brickmaking enterprises as if they comprised one and the same inseparable sector. Some small-scale artisanal brickmakers may fit the enterprise model in that their business can develop technologically, managerially and financially. Others are what may be referred to as subsistence operations, however. Such operations are not phenomena unique to the majority world. In my homeland, Wales, for example, Dylan Jones Evans laments the lack of ambition of 'lifestyle' businesses (Jones Evans, 2001). Subsistence or lifestyle business may have either no capacity to grow or no desire to do so. While the capacity for growth may be developed in some cases, I suggest that other businesses have largely reached the size that is compatible with the context in which they operate. This does not mean that they are unable to innovate to survive. As, for example, fuelwood becomes scarce and expensive and a small-scale brickmaking operation is threatened, borrowing, investing and growing the business is only one strategy. Brickmakers should also be able innovate to increase efficiency or find alternative fuels while retaining their characteristic scale.

Artisanal brickmaking clusters vary in scale. In some African countries, for instance, a cluster may be the workplace of tens of people. In India this number might be hundreds, and in China thousands. Overall, however, the numbers of people employed in brickmaking worldwide are small relative to other sectors of the global economy. It could therefore be argued that there would be little impact on national economies if most or even all small-scale brickmakers were forced to close down due to economic pressures. This macro view ignores the underlying importance of small- and medium-scale brickmaking in many situations. There are many issues that must be considered and these slip under marco-economic dragnets such as GDP per capita. Paying attention to such issues is fundamental to 'economics as if people mattered' (Schumacher, 1973).

There are alternatives to fired clay bricks for housing, communal and commercial buildings. These include concrete blocks, stabilized soil blocks, timber and stone. Those consumers of building materials who have the luxury of choice are, then, less likely to be affected by the knock-on effect of any crisis in fuel supply than are small-scale brickmakers themselves. In many countries, unemployment, underemployment and employment insecurity are rife. In such locations, if small-scale brickmakers did lose their livelihoods, they would not simply walk into jobs in other sectors. Many brickmakers have been producing bricks all of their working lives. They do not have the skills to obtain work or exploit business opportunities in other fields where the demands are rapidly changing. In the globalized world economy, labour is not a valued commodity. Though it may not noticeably affect GDP, the everyday impact of increased numbers of unemployed people on societies is severe. If brickmakers from rural or peri-urban areas are forced to migrate into towns and cities in search of work, they will most likely swell the ranks of not only the unemployed but also the homeless or ill-housed, perhaps even the mendicant or criminal.

If a fuel crisis did put small-scale brickmakers out of business, formal sector medium- and large-scale brickmakers will inevitably fill part of the gap in supply. These larger-scale brickmakers would generally produce bricks to a higher specification and sell them at a higher price, though, and such bricks would not suit everyone. Apart from being unaffordable, many people do not require bricks of the quality or in the quantity that larger firms supply. Artisanal brickmaking operations are flexible enough to meet local needs. People who have no choice and who rely on low-cost bricks to construct their homes and buildings, typically some of the poorest in the community, would be the ones to suffer if small-scale brickmakers were squeezed out. In short, without small-scale brickmakers the shelter crisis and poverty would be bound to worsen even if that change did not register on the scale of GDP per capita.

So, subsistence businesses can be regarded as the poorest supplying the needs of the poorest. As we saw when we reviewed the shelter crisis in Chapter 1, it does not look as if the numbers of the poorest are set to decline significantly in the foreseeable future. Some small-scale brickmakers will certainly migrate out of poverty via enterprise development. But life is more complex than the one-size-fits-all model of development as growth would have it. Another segment of the brickmaking sector will remain suppliers of lesser quality and cheaper building materials, if for no other reason than that a large market niche will continue to exist or even expand. (Here is a case of intrinsic value applying to humans as part of nature. All people are valuable to humanity but not all are valued by us when we apply crude financial measures.) Small-scale brickmakers can only remain suppliers to the poorest if they can obtain fuel, though there is the alternative of producing walling materials that do not require fuel enrgy, such as stabilized soil blocks, a technology with Practical Action is also involved.

Thus, any meaningful attempt to address the shelter crisis must surely include assistance to small-scale building materials producers.

In livelihood terms, we have argued that both growth and subsistence brickmaking operations need to innovate on energy efficiency and fuel substitution. Growth is only an option for a limited number of enterprises, however. Others cannot follow the route of formalization and investment as a development strategy. This is where appropriate technology comes in. We must ask ourselves an important question. If we can technically substitute wastes for primary fuels, it is environmentally sustainable to do so, *and* there is a general livelihood imperative, what is preventing mass-scale adoption of the technology in small-scale brickmaking? In tune with our approach so far, I suggest some parts of the answer to this question may be general and others specific to a particular instance. In general, we seem to have technologies that *are* appropriate, 'building on small-scale, low-cost, environmentally friendly and non-violent local knowledge and skills via a dynamic and participative process', as we noted in Chapter 1. This is evidently not a sufficient condition to ensure mass-scale dissemination and adoption, however. Perhaps if we focus on specific cases, we will find clues as to why not. So, let us continue to focus on the wastes we considered in the previous section, being careful to draw on other examples from our case studies and reviews where they can be enlightening.

Despite our claim of general technical feasibility, the limit on incorporating rice husks or sawdust into bricks as fuels in Peru does appear to be largely technical. Only a certain percentage can be incorporated before the properties of green and fired bricks are critically affected. Brickmakers generally adopt the technology up to this technical limit and there is reportedly no problem with either the supply or price of rice husks or sawdust. Unlike some agricultural residues, these agro-industrial by-products are available in sufficient density at primary processing sites, i.e. rice mills and sawmills. With the technically successful technology of hand-moulded briquettes of coal-dust and clay, by contrast, the limit on adoption is economic. Where there is a ready supply of coal-dust and the cost of making briquettes is comparable to fuelwood, brickmakers use briquettes in combination with coal-dust in layers and an oil burner to ignite the kiln. It is the saving in time and labour that dictates the continuation of this combination-fuel technology. The limiting factor on adoption of the waste oil burner is not, as would be rational, the relatively high cost of the technology, a cost that includes the diesel for the motor that pumps the waste oil through the burner. Brickmakers accept the increased cost not only because of the time and labour saved but also, it seems, because the waste oil burner is perceived as modern. A limiting factor is apparently shortage of supply due to competing demands for the waste oil.

While our initial appraisal suggests the limit on the use of wastes in Peru could be as much technical as economic, the relatively cheap and ready supply of fuelwood in most regions may well be inhibiting technology development. Despite the acknowledged environmental problems and concomitant legislation, in the final analysis most brickmakers *can* get hold of fuelwood, whether legally, quasi-legally or wholly illegally. Maybe an economist would judge that this fuelwood is not expensive enough to act as a spur to further innovation? Grinding rice husk for example, would allow more fuel to be incorporated into bricks. Moreover, low-pressure briquetting of wastes such as ground rice husks and sawdust in combination with a binder could increase fuel options. Grassroots economics means that innovation requires more than a spur. Whether or not fuelwood remains relatively cheap, small-scale brickmakers will not, in the majority of cases, be able to access the capital to buy machines such as grinding equipment or a block press. They would not, furthermore, be able to bear the costs of technology development: the inevitable failures involved as limits are pushed.

Let us now consider Zimbabwe. Boiler ash has been used as an auxiliary fuel, supplementing both fuelwood and coal in clamps or scove kilns. It has been successfully deployed between layers of bricks and the finer fraction has been incorporated into bricks as a fuel. Technology is not a limiting factor. Fuelwood is scarce and coal is relatively expensive in Zimbabwe, so the substitution of waste for a proportion of these primary fuels should be economically attractive. There could be a problem with competition as others apart from brickmakers recognize that the ash has a calorific and hence monetary value. Furthermore, the supply chain is difficult to establish, with power stations and tobacco farms obviously not treating boiler ash as their principal commercial product. Typically, they do not take orders, control stock or make deliveries, for example. In the current crisis, there is frequently no fuel for vehicles and hence no transport that brickmakers can hire to collect any sort of fuel. Such factors combine to make the availability of boiler ash a problem. It is not anyway certain that, in the long term, a sufficient quantity of ash will be available to support brickmaking across the nation. Moreover, now that the ash has a price and some sort of market for it has been established, small-scale brickmakers must find more money upfront, whereas fuelwood can still be (mis)appropriated or obtained on credit from their familiars operating in the same informal sector.

Zimbabwe's economy is currently in such a mess that it is difficult to say whether using boiler ash in brickmaking remains viable. Is there, in fact, an economic rationale that can be applied to countries like Zimbabwe? I have argued that a portion of most societies, though not most nations, will remain outside the prescriptions of globalization and development as economic growth. In other words, many countries will de facto have two economies, the one frantically modernizing, desperate for growth and increasingly defined by consumerism, the other characterized by at worst subsistence and at best sufficiency and frugality. I have also argued, however, that environmental sustainability and livelihoods demand that small-scale brickmakers, at least, share an imperative to innovate that runs through both economies. But this two-tier model of development with a common imperative cannot be applied to nations ravaged by continual strife. Sustainable development demands, in the first instance, peace and respect for human rights. In this, Zimbabwe is patently beyond the pale and also beyond our analytical ability. For brickmakers there, as for the majority of the population, economics is currently a matter of survival in the most immediate term.

The use of bagasse blocks as a substitute for fuelwood in Sudan is technically feasible, environmentally sustainable and desirable in terms of the livelihoods of brickmakers and the shelter needs of their customers. Furthermore, there are almost literally mountains of the stuff available and the supply appears secure. The problems with dissemination on the national scale are seemingly economic. The relatively high cost and scarcity of reportedly the most technically suitable binder for blocks, molasses, combined with the high cost of road transport and a lack of disseminated information, undermined attempts to introduce the technology on a mass scale. As we noted in Chapter 5, this is the most obvious example from all of our case studies of an economic system that works against ecological modernization.

To a degree, the economists are right: fuelwood is not expensive enough to spur innovation. To ensure intergenerational equity, the price of fuelwood should take into account the cost to future generations of deforestation and net carbon dioxide emissions. Alternatively, burning wastes that have less negative environmental impacts should be subsidized. Unfortunately, economics is not only a dismal science, as the historian Thomas Carlyle deemed, it is also a myopic one. In their analyses, economists find it impossible to adopt the very long-term perspective that sustainable development demands. To be fair, this is hardly surprising. For what will an adequately functioning greenhouse effect be worth to our great-grandchildren translated into cash terms today? How much should fuelwood cost in Sudan in order that the competition with wastes-as-fuels is truly fair?

Conclusion: what's to be done?

I should begin by stating that I do not want this conclusion to sound like either an abstract wish-list or a set of impracticable prescriptions for social change. These alternatives may actually amount to much the same thing in substance, differing only in tone between entreaty and demand. It is as hard not to plead on behalf of the increasing numbers of the global poor as it is not to insist that the system must change to eradicate poverty. So, should I entreat more development aid to continue work such as that done by the organization Practical Action? Or should I demand that the international community of governments regulate the development playing field in favour of sustainable development? While I suppose I have very slightly more chance of achieving the former, experience suggests that at best it could amount to only small changes for relatively few people. So, let me beard the lion in its den.

I believe we have shown that for the use of wastes as fuels by small-scale brickmakers to be viable, as well as feasible and sustainable, two aspects need particular attention. First, more technology development is needed. Second, we need the reflexive institutional reorganization of society that is fundamental to ecological modernization (Mol, 1995). A central focus of ecological modernization 'is the decoupling of economic growth and environmental degradation', thereby enabling so-called green growth (Revell and Rutherford, 2003). In other words, governments should enact and implement necessarily radical policies in favour of sustainable development. Such policies would include, for example, strictly enforcing regulation on deforestation and setting carbon emission limits far in excess of the modest targets of the Kyoto Protocol. This second aspect of change is by far the most critical and can be viewed as almost automatically enabling the first, i.e. promoting and supporting the development of sustainable technologies would be government policy and practice. While I am certain governments around the world would protest that they *are* considering sustainable development in policymaking, I contest that rather than radical institutional change they view ecological modernizsation as a political programme of, at best, modest socio-technical reform. Furthermore, I am convinced that reform, and certainly the current scale and speed of reform, is totally inadequate to the task.

Even the radical institutional reorganization and technological transformation of society that is ecological modernization would not be sufficient to ensure the livelihoods of *all* small-scale brickmakers. It would, I believe, help that segment of the sector with the capacity and scope for enterprise development. However, I have proposed that another segment of the small-scale brickmaking sector is intimately concerned with meeting the basic needs of the poorest in society and cannot grow away from that community. There is not a rigid boundary between subsistence and growth enterprises and individual operations will migrate both ways, On the sectoral scale, however, I am claiming that subsistence or sufficiency operations will remain part of society in the long term. For the foreseeable future, I am assuming that these producers will be needed because poverty is not going to disappear in the blink of an eye regardless of changes to institutional frameworks. In any event, I reject the idea that all socio-economic problems can or will be solved by growth. In at least the medium to long term, then, another approach must go hand in hand with ecological modernization, namely sufficiency.

> Swadeshi [meaning, essentially, local self-suffiency] avoids economic dependence on external market forces that could make the village community vulnerable. It also avoids unnecessary, unhealthy, wasteful, and therefore environmentally destructive transportation. The village must build a strong economic base to satisfy most of its needs, and all members of the village community should give priority to local goods and services. (Satish Kumar in Goldsmith and Mander, 2001)

Though theorists may view ecological modernization and self-sufficiency as antithetic (Murphy, 2001), in reality we always and inevitably live with contradictions in society. Practice does not change so easily and consistently as the enveloping logic of some grand theories tends to suggest. For the world's poor, particularly, everyday needs must be met regardless of 'the sound of ideologies clashing', as the singer Billy Bragg put it. Though, like Sachs in the introductory quote to Chapter 1, noble proponents of sufficiency, such as Mahatma Gandhi, Fritz Schumacher and Nelson Mandela, view it as an national-scale alternative to, say, export-oriented capitalism, I am suggesting that the two systems will exist side by side. In fact, they already do. Vandana Shiva identifies three economies, in fact:

> As the dominant economy myopically focuses on the working of the market, it ignores both nature's economy and the sustenance economy, on which it depends. In a focus on the financial bottom line, the market makes invisible nature's economy and people's sustenance economies... In the sustenance economy, people work to directly provide the conditions necessary to maintain their lives. (Shiva, 2005)

For the self-sufficient Gandhian village community expounded by Satish Kumar read local rural, peri-urban and urban communities not on the guest list for the neo-liberal globalization party. The imperative for politicians and policymakers is to recognize the simultaneous existence, and right to existence, of two economic systems, two communities. They should not prescribe a single, growth-oriented path out of poverty. As Rod Aspinwall said in the quote at the beginning of this chapter, sustainability is 'a place where there's an economy for all'. So, apart from favouring ecological modernization, regulation must also protect communities by nurturing sufficiency.

Though it is quite clear what is expected of politicians and policymakers, the current climate of global neo-liberalism makes acceptance of my suggestions not impracticable but ideologically unpalatable. My notion of the sustainability that would enable the mass-scale use of wastes as fuel in brickmaking is, no doubt, idealistic. As a vision, though, I concur with Rod Aspinwall, it is worth upholding. NGOs and individuals must continue to lobby for practical solutions to poverty, pragmatic solutions that circumvent ideological dogma, solutions that match the needs of particular communities, and solutions that are environmentally sustainable. We must be very clear about what we mean by sustainability and continue to develop mechanisms to operationalize the concept. On the ground, in the direct context of this book, we must continue to work with all small-scale brickmakers to develop technologies that help ensure both their livelihoods and also the supply of affordable building materials to the increasing number of people who have no other choice.

APPENDIX
Photocopiable forms

Environmental impact of national projects

Designation of environmental effects:
A=significant, B=should be examined, C=of minor significance,
D=insignificant

Is the proposal believed to cause a change in or effect:	*A*	*B*	*C*	*D*

1. WATER

   1.1 Surface water

   - Discharges of organic substances, including toxic substances, into lakes & water courses?
   - Discharge into coastal areas or marine waters?
   - Quantity of surface water or water level?
   - Quality of salt water or freshwater?
   - Natural ecosystems & habitats in salt or fresh water?
   - Drinking water supply or reserves?
   - Consumption/withdrawal of water?

   1.2 Groundwater

   - Percolation to groundwater?
   - Groundwater quality?
   - Quantity of groundwater?
   - Drinking water supply or reserves?
   - Consumption/withdrawal of water?

2. AIR

   - Emissions into air?
   - Air quality (e.g. acid gases, particulate or toxic substances)?
   - Obnoxious smells
   - Changes in precipitation quality?

3. CLIMATE

   - Emissions of greenhouse gases?

- Other factors, including deforestation, which may cause local or global changes in climate?

4. THE EARTH'S SURFACE & SOIL
   - Applicability or cultivation value of soil?
   - Percolation or accumulation of toxic or hazardous substances in the soil?
   - Water or wind erosion?
   - Soil in the case of changes in groundwater level?
   - The structure of the strata?
5. FLORA & FAUNA, INCLUDING HABITATS & BIODIVERSITY
   - The number of wild plants or animals of any species or the distribution pattern of species?
   - The number or distribution pattern of rare or endangered species?
   - Import or export of new species, including genetically modified organisms?
   - Quality or quantity of habitats for fish & wildlife?
   - Structure of function of natural ecosystems?
   - Vulnerable natural or uncultivated areas (e.g. bogs, heaths, uncultivated dry meadows, salt marches, swaps and coastal meadow, watercourses, lakes, humid permanent grasslands and coasts)?
   - The reproduction or natural patterns of movement or migration of fish & wildlife species?
   - Cultivation methods or land use in the agricultural or forestry sectors?
   - Fisheries, catches or the methods applied in deep-sea or freshwater fishing?
   - Open-air activities or traffic in the countryside which may affect the flora & fauna or cause wear & tear on the vegetation?
6. LANDSCAPES
   - The total area or the land use within areas used
   - Geological processes such as soil migration and water erosion?
   - Geological structures in the landscape, e.g. river valleys, ridges & coastal structures?
   - Permanent restrictions on land use which reduce the future possibilities of use of the open land?
   - The extent or appearance of archaeological or historical sites, or other material assets?
7. OTHER RESOURCES
   - Cultivation, cutting, catching or use of renewable resources, e.g. trees, fish or wildlife?

- Exploitation or use of non-renewable resources such as fossil fuels, minerals, raw material (sand, clay)?

8. WASTE
   - Wastes, residues or quantities of waste disposed of, incinerated, destroyed or recycled?
   - Treatment of waste or its application on land?
9. HISTORICAL BUILDINGS
   - Buildings with architectural, cultural or historical value and with possibilities of preservation and restoration?
   - Buildings or historical monuments which require repair because of a change of the groundwater level or air pollution?
10. PUBLIC HEALTH & WELL-BEING
    - Acute &/or long-term health risk in connection with food, drinking water, soil, air, noise, or handling of hazardous or toxic substances?
    - Risk associated with exposure to noise?
    - Recreational experiences & facilities, including changes in the physical appearance of landscapes, natural or uncultivated areas?
    - The function & environment of towns, including green areas & recreational facilities?
    - Aesthetic values or visual experiences (e.g. scenery, urban environment or monuments)?
11. PRODUCTION, HANDLING OR TRANSPORT OF HAZARDOUS OR TOXIC SUBSTANCES
    - Risk of fire, explosions, breakdowns or accidents & emissions?
    - Risk of leaks of environmentally alien or genetically engineered organisms?
    - Risks associated with electromagnetic fields?
    - Risk of radioactive leaks?
    - Risk of breakdowns or accidents during transport of substances of materials?
    - Other effects related to the security and safety of the population (e.g. traffic accidents, leaks)?

Energy monitoring form

Energy consumption of brick firing process

Name of Producer	Location/Address	Dates of Firing
Type of Clamp/Kiln	Type(s) of Fuel	Mass of Fuel(s) Used (kg)
Calorific Value(s) (kJ/kg)	No. Of Green Bricks	Avg. Mass Of Bricks (kg) Wet = Dry-green = Fired =
Brick Moisture Content	Method Of Forming	Weather Conditions

Calculation Of Kiln Efficiency	Qualifying Information
Mass of green brick =	
Total moisture content =	(i) Vitrification temp =
Drying energy =	
Wood energy =	(ii) Max kiln temp =
Coal energy =	
Gross energy =	(iii) Firing time =
Firing energy =	
Mass of fired brick =	
Specific firing energy =	

COMMENTS

NAME, CONTACT DETAILS & DATE

References

APGEP-SENREM/ITDG (2002) Utilization of rice husks as a source of fuel for brickmakers. APGEP-SENREM/ITDG, Lima.

Arler, F. (2001) 'Distributive Justice and Sustainable Development', in *Our Fragile World: Challenges and opportunities for sustainable development*, ed. M.K. Tolba, EOLSS/UNESCO, Oxford.

Bairiak, J.T. (1997) *Improving Bricks' Properties and Substituting Fuelwood with Oil*, ITDG Sudan.

Bairiak , J.T. (1998) *Utilisation of Cow Dung and Bagasse in Brickmaking. Case studies on the use of wastes*, ITDG Sudan/GTZ.

BBC (2005) [accessed 11 January 2006] *Country profile: Zimbabwe*. [Online] http://news.bbc.co.uk/1/hi/world/africa/country_profiles/1064589.stm

BBC (2005) [accessed 3 January 2006] *Clean Coal Technology: How it works*. [Online] http://news.bbc.co.uk/2/hi/science/nature/4468076.stm

BDA (2000) *A Sustainability Strategy for the Brick Industry*. Brick Development Association, Windsor.

BDA (2001) *Observations on the use of Reclaimed Clay Bricks. Properties of Bricks and Mortars Generally*. Brick Development Association, Windsor.

Blackman, A. (2000) 'Making Small Beautiful - Lessons from Mexican Leather Tanneries and Brick Kilns', *Small Enterprise Developme*nt, 11 (2), June, pp. 4–14.

Boulding, K.E. (1966) 'The Economics of the Coming Spaceship Earth' in *Environmental Quality in a Growing Economy, Essays from the Sixth RFF Forum*, ed. H. Jarrett, Resources for the Future/Johns Hopkins University Press, Baltimore.

Bullen, A. and Mark Whitehead (2005) 'Negotiating the Networks of Space, Time and Substance: a geographical perspective on the sustainable citizen'', *Citizenship Studies*, 19(5), pp. 499–516.

Callicott, J.B. (1994) *Earth's Insights*, University of California Press, Berkley.

Campbell, C. (2005) *Oil Crisis*. Multi Science Publishing Company, Brentwood.

Campbell, W.P. and Will Pryce (2003) *Brick: A World History*, Thames & Hudson Ltd., London.

CSS (2004) *Sudan in Figures 1999–2003*, Central Statistics Office, Government of Sudan, Khartoum.

Croall, S. and W. Rankin (2000) *Environmental Politics*, Icon Books, Cambridge.

Christensen, P., Lone Kørnøv, and Eskild Holm Nielsen (2003a) *The Outcome of EIA in Denmark*, Ministry of Environment/Aalborg University, Copenhagen.

Christensen, P., Lone Kørnøv, and Eskild Holm Nielsen (2003b) 'Mission Impossible: Does EIA secure a holistic approach to the environment', *International Association for Impact Assessment*, Morocco.

Cloke, P. and O. Jones (2004) 'Turning in the Graveyard: Trees and the hybrid geographies of dwelling, monitoring and resistance in a Bristol cemeter', *Cultural Geographies*, 11, pp. 313–41.

Commoner, B. (1971) *The Closing Circle*, Alfred A. Knopf Inc., New York.

Daly, H.E. (1992) *Steady-state Economics*, Earthscan Publications, London.

Daly, H.E. (1996) *Beyond Growth: The Economics of Sustainable Development*, Beacon Press, Boston.

Daly, H.E., J.B. Cobb, and C.W. Cobb (1994) *For the Common Good: Redirecting the economy toward community, the environment, and a sustainable future*, Beacon Press, Boston.

Deffeyes, K.S. (2005) *Beyond Oil: the view from Hubbert's Peak*, Hill & Wang, New York.

Denmark (1995) *Guidance on Procedures for Environment Assessment of Bills and Other Government Proposal*. Danish Prime Minister's Office, Copenhagen.

Dondi, M., M. Marsigli, and B. Fabbri (1997) 'Recycling of Industrial and Urban Wastes in Brick Production: A review (Part 1)', *Tile & Brick International*, 13 (3), pp. 218–25.

European Commission (2003) *The New SME Definition: User guide and model declaration*, Enterprise and Industry Publications, Luxemborg.

Ekins, P. (1999) *Economic Growth & Environmental Sustainability: The prospects for green growth*, Routledge, Oxford.

Florman, S.C. (1994) *The Existential Pleasure of Engineering*, St Martins Press, New York.

George, S. (2004) *Another World Is Possible If...* Verso, London.

Goldsmith, Edward and Jerry Mander, eds. (2001) *The Case Against The Global Economy: And for local self-reliance*, Earthscan, London.

Haleja, R.B. *et al.* (1985) 'Agro-Industrial Wastes in Brickmaking', *Building Research Practice*, 18 (4), July/August, pp. 248–52.

Hamid, M.H. (1994) 'Wood Use in Brickmaking Industry in Sudan, *Forest Product Consumption Survey*, FNC/FAO, Khartoum.

Hood, A.H. (1999a) *Financial Evaluation of Shambob Bricks Production Co-operative, Takroof, Kassala*, ITDG Sudan.

Hood, A.H. (1999b) *Scotch Kiln Construction, Firing and Energy Efficiency*, ITDG Sudan.

Hood, A.H. (1999c) *Training SBPC Member on Improved Brick Production Techniques*, ITDG Sudan.

Jones Evans, Dylan (2001) *Creating An Entrepreneurial Wales (Gregynog Papers)*, Institute of Welsh Affairs, Cardiff.

Lardinois, I. and A. van de Klundert (1993) *Organic Waste: Options for Small-scale Resource Recovery, Urban Solid Waste Series 1*, TOOL Amsterdam and Waste Consultants Gouda.

Latour, B. (1993) *We Have Never Been Modern*, Harvard University Press, Cambridge, Mass.

Leaky, R.R.L. (1996) *The Sixth Extinction: Biodiversity and its survival*, Phoenix, London.

Leopold, A. (1949) *A Sand Country Almanac and Sketches Here and There*, Oxford University Press, New York.

Mason, K. (1999) *The Small-Scale Vertical Shaft Lime Kiln: A practical guide to design, construction and operation*, ITDG Publishing, Rugby.

Mason, K. (2000a) *Assessing the Technical Problems of Brick Production: A guide for brickmakers and field-workers*, ITDG Publishing, Rugby.

Mason, K. (2000b) *Ten Rules for Energy Efficient, Cost Effective Brick Firing: A guide for brickmakers and field-workers*, ITDG Publishing, Rugby.

Mason, K. (2001) *Brick By Brick: Participatory technology development in brickmaking*, ITDG Publishing, Rugby.

Mason, Kelvin (2005) 'Understanding Visions of the Sustainable Citizen in Wales: An analysis of theories, policies and imaginings of identity, participation, rights and duties', MA Thesis, University of Wales Aberystwyth.

McDonough, W., and M. Braungart (2002) *Cradle to Cradle: Remaking the way we make things*, North Point Press, New York.

Mayorga, E. (1999) *Efficient use of energy in small-scale brick production.*

Merschmeyer, G. (2003) *Utilization of Agricultural Wastes in Brick Production 1: Firing of clay bricks and tiles with rice husks in periodically built clamps in Tanzania*. basin/GATE/GTZ.

Milligan, Spike (1973) *Small Dreams of a Scorpion*, Penguin, Harmondsworth.

Ministry of Lands Annual report for 2004, Ministry of Lands, Harare

Moffat I. et al. (2001) *Sustainable Prosperity: Measuring resource efficiency*, Sustainable Development Unit of Department of Environment, Transport and the Regions, UK.

Mol, A.P.J. (1995) *The Refinement of Production: Ecological modernisation theory and the chemical industry*, International Books, Utrecht.

Muir, J. (1992) *John Muir: The eight wilderness-discovery books*, Diadem Books, Spean Bridge.

Murphy, J. (2001) *Ecological modernisation: The Environment and the Transformation of Society*, OCEES Research Paper No. 20, Oxford Centre for the Environment, Ethics and Society, Mansfield College, University of Oxford.

Palmer, J.A., ed. (2001) *Fifty Key Thinkers on Environment*, Routledge, Oxford.

Pineiro, M. (2005) *Cleaner brick making processes in Arequipa and Cusco: survey of the situation*. Lima.

Raut, A.K. (2002) ('Brick Kilns on Health', *The Kathmandu Post*, Sunday, 21 April 2002. [Online] www.environmentnepal.com.np/articles_d.asp?id=58

Revell, A. and R. Rutherford (2003) 'UK Environmental Policy and the Small Firm', *Business Strategy and the Environment* 12, pp. 26–35.

Roberts, P. (2004) *The End of Oil: On the edge of a perilous new world*, Houghton Mifflin, Boston.

Russell, A. (1996) *Survey of the State of the Art of Energy Use and Efficiency in Building Material Production*, ITDG Publishing, Rugby.

Sachs, W. (1999) *Planet dialectics : explorations in environment and development*, Zed Books, London, New York.

Schumacher, E.F. (1973) *Small Is Beautiful: A study of economics as if people mattered*, Vintage, London.

Shiva, Vandana (2005) *Earth Democracy: Justice, Sustainability and Peace*, South End Press, Cambridge Mass.

Stokes, J. (2004) *The Heritage Trees of Britain and Northern Ireland*, Constable & Robinson, London.

Tawodzera, P. (1994) 'Using waste as an alternative fuel in firing bricks in Zimbabwe', Article submitted to *Basin News* for publication.

Tawodzera, P. (1998) *Use of brickmaking equipment: Zimbabwean experience*, Intermediate Technology Zimbabwe.

TERI (2000) *Resource Utilization Improvements in Brick Industry*, TERI India.

UN-HABITAT (2003) *The Challenge*, United Nations Human Settlements Programme.

WCED (1987) *Our Common Future*, Zed Books, London.

Whatmore, S. (2002) *Hybrid Geographies: natures, cultures, spaces*, Sage, London.

Wilson, R.C.L., S.A. Dury, and J.L. Chapman (1999) *The Great Ice Age: Climate change and life*, Routledge, Oxford.

Zimbabwe (2002) *Zimbabwe Census 2002, National Report*, Central Statistical Office Harare, Zimbabwe. http://data.unaids.org/pub/GlobalReport/2006/200605-FS_SubSaharanAfrica_en.pdf

Zimbabwe Government and UNDP (2004) *Zimbabwe Millennium Development Goals: 2004 progress report*.

Zimbabwe Government (2005) *Zimbabwe Millennium Development Goals, 2005 progress report*, UNDP (draft), Harare.

Further reading

Almeida, J.J.R.C. and Lucas J. A. de Carvalho (1991) 'Incorporation of Petroleum Waxes in Clay Bricks', *Industrial Ceramics*, 11 (4), pp. 194–7.

Anderson, M. (undated) 'The Clamp Firing of Flyash Bricks', Notes, Department of Ceramic Technology, North Staffordshire Polytechnic, Stoke-on-Trent, UK.

Anderson M. and G. Jackson (2004), 'The Beneficiation of Power Station Coal Ash and its Use in Heavy Clay Ceramics', *Transactions and Journal of the British Ceramics Society*, 82, March/April

Arends, G.J. and S.S. Donkersloot-Shouq (1985) *An Overview of Possible Uses of Sawdust*, Tool Foundation, Amsterdam; CICAT, Delft; and CICA CMP, Eindhoven, Netherlands.

Bairiak, J. and M. Majzoub (1999) 'Utilization of Bagasse in Brickmaking: R & D in Sudan', *Basin Building Partnerships Wall Building Technical Brief*, GATE/GTZ, Eschborn, Germany.

Barriga, Alfredo et al., (1992) *Brick and Lime Kilns in Ecuador: An example of wood fuel use in Third World small-scale industry*, Stockholm Environment Institute, Energy, Environment and Development Series No. 13.

Bhattarai, Mukesh Dev (1993) 'Paper on Industrial Contribution to Air Quality, Urban Air Quality Management Workshop (UrbAir)', Ministry of Industry, 1–3 December 1993, Kathmandu, Nepal.

Bingh, Lars Petter (2004) 'Opportunities for Utilizing Waste Biomass for Energy in Uganda', Master's Thesis, Department of Energy and Process Engineering, Faculty of Engineering Science and Technology, The Norwegian University of Science and Technology, Trondheim.

Border Affairs Division, Texas Commission on Environmental Quality (TCEQ) [accessed 22 March 2006] *A Study of Brick-making Processes along the Texas Portion of the U.S.* (Mexican Border: Senate Bill 749, TCEQ, Austin, Texas, Document SFR-081/02, December 2002. [Online] www.tceq.state.tx.us/assets/public/comm_exec/pubs/sfr/081.pdf

Bozadjief, L. et al. (1991) 'Use of Flotation Copper Ore Tailings in Tile and Brick Bodies', *Tile & Brick International*, 7 (6), pp. 426 and 429.

Bozadjief, L. and T. Dimova (1995) 'Use of Copper Ore Dressing Slime in Brick Manufacture', *Tile & Brick International*, 11 (2), pp. 88–9.

Caligaris R. E. et al (1990) 'Bricks from Coal Tailings', *Tile & Brick International*, 6 (4), pp. 41–2.

Carter G.W., M.A. Connor, and D.S. Mansell (1982) 'Properties of Bricks Incorporating Unground Rice Husks', *Building and Environment*, 17 (4), pp. 285–91.

Central Electricity Generating Board (1967) *PFA Data Book: Bricks and other structural ceramics*.

Churchill, W.M. (1994) 'Aspects of Sewage Sludge Utilisation and its Impact on Brickmaking', *British Ceramic Transactions*, 93 (4), pp. 161–4.

De Guiterrez, R.M. and S. Delvasto (1994) 'Use of Rice Husk in Ceramic Bricks', in *Ceramics - Charting the Future, Proceedings of the World Ceramics Congress, part of the 8th CIMTEC (World Ceramics Congress and Forum on New Materials, Florence, Italy, June 28 – July 4, 1994)*, ed. P. Vincenzini, Part A, pp. 255–62.

Dondi, M., M. Marsigli, and B. Fabbri (1997) 'Recycling of Industrial and Urban Wastes in Brick Production (A Review - Part 2)', *Tile & Brick International*, 13 (4), pp. 302–15.

Elwan, M.M. and M.S. Hassan (1998) 'Recycling of some Egyptian Industrial Solid Wastes in Clay Bricks', *Industrial Ceramics*, 18 (1), pp. 1–6.

Frijns, Jos (1999) 'Small-scale Industry and Cleaner Production Strategies', *World Development*, 6, pp. 967–83.

Giugliano, M. and A. Paggi (1985) 'Use of Tannery Sludge in Brick Production', *Waste Management & Research*, 3, pp. 361–8.

Harvey Reid [accessed 29 January 2003] *Development of a Low Cost, Heat Resistant, Kiln Brick, for Production of Purifiers, Kiln Firing.* [Online] http://www.purifier.com.np/kiln29-1-3.html.

Hauck, D., et al. (1990) 'Additives for Product Improvement and Reduction of the Final Firing Temperature in Brick Firing', *ZI Annual*, pp. 95–116.

Hofer, Michael (1994) 'The Exhaust Gas Problems in the Brick and Tile Industry - Regenerative afterburning at Leipfinger & Bader', *Ziegelindustrie International*, September, pp. 552–7.

Knirsch, M. et al. (1998) 'Application of Brewery Wastes in the Production of Bricks', *Tile & Brick International*, 14 (2), pp. 93–101.

Koopmans, Auke and Stephen Joseph (1993) 'Status and Development Issues of the Brick Industry in Asia, Food and Agricultural Organization of the United Nations', Regional Wood Energy Development Programme in Asia, GCP/RAS/131/NET, Field Document No. 35 (April), Bangkok.

Merienne, J. (1995), 'Energy Saving in the Brick-making Industry with the Addition of Heat-Producing Waste', *Tile & Brick International*, 11 (2), pp. 112–15.

Merschmeyer, Gerhardt (1989) *Basic Know How for the Making of Burnt Bricks and Tiles: Handbook for village brickmakers in Africa*, Misereor, Aachen, Germany.

Merschmeyer, Gerhardt (2003) *Utilization of Agricultural Wastes in Brick Production - 2: Firing of clay bricks and tiles with coffee husks in permanent built kilns in Uganda*, Basin / GATE / GTZ, Eschborn Germany. [Online]
www2.gtz.de/Basin/gate/cs26/ABT_casestudy_husks02_uganda.pdf

Mukhopadhyay T.K. and K.N. Maiti (1995) 'Utilisation of Cinder for Manufacturing Unglazed Vitrified Floor Tiles', *Tile & Brick International*, 11 (4), pp. 264–70.

Narayana, M., N.K. Edirisinghe, and W.R. Sovis (2004) 'Improvement of Combustion Efficiency of Tile Furnaces of Tile Industry by Introducing Particulate Biomass Fuels', Project Report, Renewable Energy Department, National Engineering Research and Development Centre of Sri Lanka, Ekala, Ja-Ela, Sri Lanka.

NSTA-KKK Appropriate Technology Program, National Science and Technology Authority (1981) *Rice Hull Ash (RHA)- Clay Bricks & Roof Tiles, NSDB Appropriate Technology Series*, NSTA, Manila, the Philippines.

Okongwu, David A. (1988) 'Effects of Additives on the Burnt Properties of Clay Brick', *Am. Ceram. Soc. Bull.*, 67 (8), pp. 1409–11.

Pepplinghouse, H.J. (1980) 'Utilization of Rice Hulls in Brickmaking (An Industrial Trial)', *Journal of the Australian Ceramic Society*, 16 (2), November, pp. 26–8.

PROGRAMA AGREP - SENREM / CONVENIO USAID (2001) *Conam, Proyecto Piloto Demostrativo Ambiental Utilizacion de la Cascarilla de Arroz Como Fuente Energetica en Ladrilleras*, ITDG Peru, Lima.

Slim, J.A. and R.W. Wakefield (1991) 'The Utilisation of Sewage Sludge in the Manufacture of Clay Bricks', *Water SA*, 17 (3), July, pp. 197–201.

Siefke, C. (1997) 'Production of Ceramic Building Materials in Russia from Coal Tailings', *Tile & Brick International*, 13 (20), pp. 116–31.

Stanaitis, V., et al. (1995) 'The Use of Waste from the Metalworking Industry in Ceramic Products', *Tile & Brick International*, 11 (6), pp. 450–2.

Tenaglia, A., C. Palmonari, and G. Timellini (1984) 'Ceramic Sludges as Raw Materials for Heavy Clay Products', *Interceram*, 2, pp. 31–5.

Wiebusch, B. and C.F. Seyfried (1997) 'Utilization of Sewage Sludge Ashes in the Brick and Tile Industry', *Wat. Sci. Tech.*, 36 (11), pp. 251–8.

Zani A., A. Tenaglia and A. Panigada (YEAR) 'Re-use of Papermaking Sludge in Brick Production', *Tile & Brick International*, 12/90, pp. 682–90 (Part 1) and 1/91, pp. 13–16 (Part 2).

Index